Applied Categorical Data Analysis

STATISTICS: Textbooks and Monographs

A SERIES EDITED BY

- **Vol. 1:** The Generalized Jackknife Statistic, *H. L. Gray and W. R. Schucany*
- **Vol. 2:** Multivariate Analysis, *Anant M. Kshirsagar*
- **Vol. 3:** Statistics and Society, *Walter T. Federer*
- **Vol. 4:** Multivariate Analysis: A Selected and Abstracted Bibliography, 1957-1972, *Kocherlakota Subrahmaniam and Kathleen Subrahmaniam* (out of print)
- **Vol. 5:** Design of Experiments: A Realistic Approach, *Virgil L. Anderson and Robert A. McLean*
- **Vol. 6:** Statistical and Mathematical Aspects of Pollution Problems, *John W. Pratt*
- **Vol. 7:** Introduction to Probability and Statistics (in two parts), Part I: Probability; Part II: Statistics, *Narayan C. Giri*
- **Vol. 8:** Statistical Theory of the Analysis of Experimental Designs, *J. Ogawa*
- **Vol. 9:** Statistical Techniques in Simulation (in two parts), *Jack P. C. Kleijnen*
- **Vol. 10:** Data Quality Control and Editing, *Joseph I. Naus* (out of print)
- **Vol. 11:** Cost of Living Index Numbers: Practice, Precision, and Theory, *Kali S. Banerjee*
- **Vol. 12:** Weighing Designs: For Chemistry, Medicine, Economics, Operations Research, Statistics, *Kali S. Banerjee*
- **Vol. 13:** The Search for Oil: Some Statistical Methods and Techniques, *edited by D. B. Owen*
- **Vol. 14:** Sample Size Choice: Charts for Experiments with Linear Models, *Robert E. Odeh and Martin Fox*
- **Vol. 15:** Statistical Methods for Engineers and Scientists, *Robert M. Bethea, Benjamin S. Duran, and Thomas L. Boullion*
- **Vol. 16:** Statistical Quality Control Methods, *Irving W. Burr*
- **Vol. 17:** On the History of Statistics and Probability, *edited by D. B. Owen*
- **Vol. 18:** Econometrics, *Peter Schmidt*
- **Vol. 19:** Sufficient Statistics: Selected Contributions, *Vasant S. Huzurbazar (edited by Anant M. Kshirsagar)*
- **Vol. 20:** Handbook of Statistical Distributions, *Jagdish K. Patel, C. H. Kapadia, and D. B. Owen*
- **Vol. 21:** Case Studies in Sample Design, *A. C. Rosander*
- **Vol. 22:** Pocket Book of Statistical Tables, *compiled by R. E. Odeh, D. B. Owen, Z. W. Birnbaum, and L. Fisher*
- **Vol. 23:** The Information in Contingency Tables, *D. V. Gokhale and Solomon Kullback*
- **Vol. 24:** Statistical Analysis of Reliability and Life-Testing Models: Theory and Methods, *Lee J. Bain*
- **Vol. 25:** Elementary Statistical Quality Control, *Irving W. Burr*
- **Vol. 26:** An Introduction to Probability and Statistics Using BASIC, *Richard A. Groeneveld*

Vol. 27: Basic Applied Statistics, *B. L. Raktoe and J. J. Hubert*
Vol. 28: A Primer in Probability, *Kathleen Subrahmaniam*
Vol. 29: Random Processes: A First Look, *R. Syski*
Vol. 30: Regression Methods: A Tool for Data Analysis, *Rudolf J. Freund and Paul D. Minton*
Vol. 31: Randomization Tests, *Eugene S. Edgington*
Vol. 32: Tables for Normal Tolerance Limits, Sampling Plans, and Screening, *Robert E. Odeh and D. B. Owen*
Vol. 33: Statistical Computing, *William J. Kennedy, Jr. and James E. Gentle*
Vol. 34: Regression Analysis and Its Application: A Data-Oriented Approach, *Richard F. Gunst and Robert L. Mason*
Vol. 35: Scientific Strategies to Save Your Life, *I. D. J. Bross*
Vol. 36: Statistics in the Pharmaceutical Industry, *edited by C. Ralph Buncher and Jia-Yeong Tsay*
Vol. 37: Sampling from a Finite Population, *J. Hajek*
Vol. 38: Statistical Modeling Techniques, *S. S. Shapiro*
Vol. 39: Statistical Theory and Inference in Research, *T. A. Bancroft and C.-P. Han*
Vol. 40: Handbook of the Normal Distribution, *Jagdish K. Patel and Campbell B. Read*
Vol. 41: Recent Advances in Regression Methods, *Hrishikesh D. Vinod and Aman Ullah*
Vol. 42: Acceptance Sampling in Quality Control, *Edward G. Schilling*
Vol. 43: The Randomized Clinical Trial and Therapeutic Decisions, *edited by Niels Tygstrup, John M. Lachin, and Erik Juhl*
Vol. 44: Regression Analysis of Survival Data in Cancer Chemotherapy, *Walter H. Carter, Jr., Galen L. Wampler, and Donald M. Stablein*
Vol. 45: A Course in Linear Models, *Anant M. Kshirsagar*
Vol. 46: Clinical Trials: Issues and Approaches, *edited by Stanley H. Shapiro and Thomas H. Louis*
Vol. 47: Statistical Analysis of DNA Sequence Data, *edited by B. S. Weir*
Vol. 48: Nonlinear Regression Modeling: A Unified Practical Approach, *David A Ratkowsky*
Vol. 49: Attribute Sampling Plans, Tables of Tests and Confidence Limits for Proportions, *Robert E. Odeh and D. B. Owen*
Vol. 50: Experimental Design, Statistical Models, and Genetic Statistics, *edited by Klaus Hinkelmann*
Vol. 51: Statistical Methods for Cancer Studies, *edited by Richard G. Cornell*
Vol. 52: Practical Statistical Sampling for Auditors, *Arthur J. Wilburn*
Vol. 53: Statistical Signal Processing, *edited by Edward J. Wegman and James G. Smith*
Vol. 54: Self-Organizing Methods in Modeling: GMDH Type Algorithms, *edited by Stanley J. Farlow*
Vol. 55: Applied Factorial and Fractional Designs, *Robert A. McLean and Virgil L. Anderson*
Vol. 56: Design of Experiments: Ranking and Selection, *edited by Thomas J. Santner and Ajit C. Tamhane*
Vol. 57: Statistical Methods for Engineers and Scientists. Second Edition, Revised and Expanded, *Robert M. Bethea, Benjamin S. Duran, and Thomas L. Boullion*
Vol. 58: Ensemble Modeling: Inference from Small-Scale Properties to Large-Scale Systems, *Alan E. Gelfand and Crayton C. Walker*
Vol. 59: Computer Modeling for Business and Industry, *Bruce L. Bowerman and Richard T. O'Connell*
Vol. 60: Bayesian Analysis of Linear Models, *Lyle D. Broemeling*
Vol. 61: Methodological Issues for Health Care Surveys, *Brenda Cox and Steven Cohen*
Vol. 62: Applied Regression Analysis and Experimental Design, *Richard J. Brook and Gregory C. Arnold*

Vol. 63: Statpal: A Statistical Package for Microcomputers – PC-DOS Version for the IBM PC and Compatibles, *Bruce J. Chalmer and David G. Whitmore*

Vol. 64: Statpal: A Statistical Package for Microcomputers – Apple Version for the II, II+, and IIe, *David G. Whitmore and Bruce J. Chalmer*

Vol. 65: Nonparametric Statistical Inference, Second Edition, Revised and Expanded, *Jean Dickinson Gibbons*

Vol. 66: Design and Analysis of Experiments, *Roger G. Petersen*

Vol. 67: Statistical Methods for Pharmaceutical Research Planning, *Sten W. Bergman and John C. Gittins*

Vol. 68: Goodness-of-Fit Techniques, *edited by Ralph B. D'Agostino and Michael A. Stephens*

Vol. 69: Statistical Methods in Discrimination Litigation, *edited by D.H. Kaye and Mikel Aickin*

Vol. 70: Truncated and Censored Samples from Normal Populations, *Helmut Schneider*

Vol. 71: Robust Inference, *M. L. Tiku, W. Y. Tan, and N. Balakrishnan*

Vol. 72: Statistical Image Processing and Graphics, *edited by Edward J. Wegman and Douglas J. DePriest*

Vol. 73: Assignment Methods in Combinatorial Data Analysis, *Lawrence J. Hubert*

Vol. 74: Econometrics and Structural Change, *Lyle D. Broemeling and Hiroki Tsurumi*

Vol. 75: Multivariate Interpretation of Clinical Laboratory Data, *Adelin Albert and Eugene K. Harris*

Vol. 76: Statistical Tools for Simulation Practitioners, *Jack P. C. Kleijnen*

Vol. 77: Randomization Tests, Second Edition, *Eugene S. Edgington*

Vol. 78: A Folio of Distributions: A Collection of Theoretical Quantile-Quantile Plots, *Edward B. Fowlkes*

Vol. 79: Applied Categorical Data Analysis, *Daniel H. Freeman, Jr.*

Vol. 80: Seemingly Unrelated Regression Equations Models : Estimation and Inference, *Virendra K. Srivastava and David E. A. Giles*

ADDITIONAL VOLUMES IN PREPARATION

Applied Categorical Data Analysis

Daniel H. Freeman, Jr.
Community and Family Medicine (Biostatistics)
Dartmouth Medical School
Hanover, New Hampshire

Marcel Dekker, Inc. New York and Basel

Library of Congress Cataloging-in-Publication Data

Freeman, Daniel H.
Applied categorical data analysis.

(Statistics: textbooks and monographs ; v. 79)
Bibliography: p.
Includes index.
1. Multivariate analysis. I. Title. II. Series:

Statistics, textbooks and monographs ; v. 79.
QA278.F74 1987 519.5'35 87-633
ISBN 0-8247-7752-5

MARCEL DEKKER, INC.
270 Madison Avenue, New York, New York 10016

Current printing (last digit):
10 9 8 7 6 5 4 3 2 1

PRINTED IN THE UNITED STATES OF AMERICA

To L. W.

Preface

Applied Categorical Data Analysis presents statistical methodologies for researchers in medicine, social science, and public health. These researchers often collect information about large numbers of subjects. The data are then analyzed by cross-classifying the subjects by two or more variables. The researchers interpret the resulting tables to illuminate some disease process or social phenomenon. This process leads to many potentially interesting associations. Some of these may be attributable to chance, others may only be apparent to the original researcher. Systematic methods for characterizing chance as an alternative explanation of a finding are therefore crucial to the research process. Similarly, if researchers follow a systematic and reproducible process in sifting through data, reported associations may be somewhat less arbitrary. For this reason statistical methodologies for cross-classified data are vitally important.

Often the cross-classification tables are based on a categorical data set. A categorical data set has one or more response variables, each of which takes on only a few distinct values. The analysis usually determines which of several factors predicts variation in the response variable or variables. The term ''response variable'' is equivalent to the term ''dependent variable.'' The term ''factor'' is equivalent to the term ''independent variable.'' The factor-response terminology is used because of analogies to analysis of variance, which are helpful in understanding various models for categorical data. In addition, many of the hypotheses about categorical data sets are in terms of independence among variables. Needless to say, referring to the independence of dependent and independent variables gets a little awkward.

This book provides researchers an introduction to some of the major techniques used in analyzing categorical data. This includes a review of probability and some common discrete distributions. Loglinear models, weighted least squares, and logistic regression are all presented. In addition, techniques for small samples and for survey samples are discussed. Most of the examples are drawn from published articles although occasionally an artificial data set is used to emphasize a particular point. For more than two variables most computations require the use of a computer. Throughout the book either SAS or BMDP software has been used. The code is included but various enhancements may have made it obsolete. The author will not accept responsibility beyond saying the code should provide a guide to the reader.

The book is based on notes from a one-semester course taught primarily to biostatistics and epidemiology students at Yale University from 1975 to 1982. However, students in sociology, psychiatry, and medicine seemed equally interested in the material. Students require only a minimal mathematical background to follow the material. It is assumed they have had an introduction to statistics that included either analysis of variance or regression analysis, preferably both. Occasionally, matrix multiplication is used and an occasional matrix inverse is employed, but nothing more complex than what is taught at the high school level is actually required. Proofs have been kept to a minimum. The exception is the collapsibility theorems which are proved based on a result by Shapiro (*American Statistician, 36,* 1982). These proofs are simple and are included because of the theorems' importance to study design.

A sabbatical at the Indian Statistical Institute, 1983–1984, provided me with the opportunity to convert these notes into the present text. The work was supported in part by an Indo-American Fellowship. I wish to thank Priscilla Adams, John Baron, Ann Eward, Jean Freeman, and Hebe Quinton for their comments on various chapters. Priscilla Adams has also patiently helped edit and index the manuscript, a task for which I am particularly grateful. I especially wish to thank Gary Koch for introducing me to categorical data analysis and trying to give me a good foundation in statistical methods.

I wish to acknowledge the editors of the *American Journal of Epidemiology*, the *American Journal of Public Health*, the *American Sociological Review*, the *Archives of General Psychiatry*, *Biometrics*, the *Journal of Chronic Disease*, the *Journal of the National Cancer Institute*, the *Journal of the Royal Statistical Society*, the *International Statistical Review*, and the *New England Journal of Medicine* for permission to reprint several tables. I wish to acknowledge the American Statistical Association, Holt, Rinehart, & Winston, Inc., the International Biometric Society, the Iowa State University Press, Lifetime Learning Publications, Little, Brown & Co., the MIT Press, the SAS Institute, and the National Center for Health Statistics for permission to reprint several tables. I also acknowledge Gary G. Koch, Adrian M. Ostfeld, and Larry Wexler, for permission to publish several previously unpublished tables.

Daniel H. Freeman, Jr.

Contents

Applied Categorical Data Analysis

1
Introduction

Health, medical, and social scientists collect and analyze data about individuals as a regular part of their research and clinical activities. Some of these data indicate the presence of diseases or other phenomena of interest. Other variables are attributes thought to be associated with the responses of interest. Finally, individuals have characteristics that may affect the associations being studied. When the variable being studied or the "response" takes on a limited number of values, the method of analysis requires the techniques that are the subject of this book.

An example might be a study of hypertension. Investigators in the early 1970s were interested in whether controlling elevated blood pressure through a carefully specified drug treatment regime was an appropriate medical recommendation for the general population. These investigators tried to answer this question by comparing a specific drug treatment regimen called *stepped care* to referral of hypertensives to a community physician. The latter is called *referred care*. The death of a participant during the five-year follow-up is a dichotomous response.

The study just described led to summary tables such as that shown in Table 1.1. Of the 3903 hypertensives receiving stepped care, 231 or 5.9 percent had died after five years. Comparing this to the 7.4 percent of referred care subjects who died, we might well conclude that the stepped care intervention should be recommended. A simple statistical test such as the Pearson chi-square, discussed in Chapter 3, would tend to support this conclusion.

The effects of drug treatments are known to depend on the baseline physical status of individuals. This suggests the planned intervention should be examined within several subgroups in order to maximize the generality of the findings. The overall results from this study seem fairly clear cut, but a thorough investigator might want to know whether the conclusion holds after adjusting for another variable, such as ''end-organ damage.'' This concern leads to examining a three-variable table which is ''stratified'' on end-organ damage (Table 1.2). People without end-organ damage had survival rates in excess of 94 percent, regardless of their treatment. When end-organ damage was present, the survival rate dropped to less than 85 percent. However, for both groups the stepped-care group had higher survival. Moreover, the relative advantage of stepped care is the same for both groups. The scientist is now confronted with the question of how to report the data. Can the simpler 2×2 table be validly reported, or is there an obligation to show the more complex table? The answer to this question depends, in part, on the collapsibility conditions found in Chapters 4 and 6.

Table 1.1 Five-Year Survival by Treatment Program

Treatment Program		Status at Five-Year Follow-up: Dead	Alive	Total
Stepped care	n	231	3672	3903
	%	5.9	94.1	100.0
Referred care	n	291	3631	3922
	%	7.4	92.6	100.0

Source: Hypertension Detection and Follow-up Program Cooperative Group (1982). Reprinted by permission of *The New England Journal of Medicine*.

Table 1.2 Five-Year Survival by Treatment Program and End-Organ Damage at Baseline

End-Organ Damage	Treatment Program		Status at Five-Year Follow-up: Dead	Alive	Total
No	Stepped care	*n*	153	3249	3402
		%	4.5	95.5	100.0
	Referred care	*n*	199	3263	3462
		%	5.7	94.3	100.0
Yes	Stepped care	*n*	78	423	501
		%	15.6	84.4	100.0
	Referred care	*n*	92	368	460
		%	20.0	80.0	100.0

Source: Hypertension Detection and Follow-up Program Cooperative Group (1982). Reprinted by permission of *The New England Journal of Medicine*.

In some situations the results are even more complex. Suppose that the study comparison of interest changes across the levels of a third variable. When the association between two (or more) variables depends on which level of a third variable is considered, the phenomenon is sometimes referred to as *interaction*. Again the techniques of categorical data analysis can be used to identify the significant relationships and generate appropriate summary tables.

1.1 OBSERVATIONS, VARIABLES, AND VALUES

The analysis of categorical data requires answering three interrelated questions:

1. Who or what is being studied?
2. What characteristics of the subjects are being studied?
3. What are the constituents of the characteristics?

The answer to the first question determines what are often referred to as the *units of analysis*. In the Hypertension Detection and Follow-up Program, the units were the individuals randomized to one of the treatment groups. In other studies the units can be more abstract. For example, counties or states might be the units of analysis for an *ecological study*. In general, the units of analysis are the simplest objects actually observed. Usually, it is about this object that we want to make inferences.

The complete collection of units is termed the *population*. Ordinarily, a study will contain only a subset or sample of units from the population. The way in which we select the sample or subset of units determines the assumptions that will permit a valid statistical analysis. We will usually assume that the sample selection is based on *simple random sampling*. There are many good texts on constructing such samples (e.g., Levy and Lemeshow 1980). Alternatively, carefully designed observational studies may make the simple random assumption appropriate.

In the hypertension example, it may be appropriate to assume that the participants were drawn as a simple random sample. Moreover, it is usually appropriate to assume that the intervention was applied at random. The first assumption will permit inferences from the particular sample to a larger population. The second assumption helps to avoid attributing the effect of the intervention to some bias of the investigators.

The answer to the second question tells us what characteristics of subjects are being studied. Those characteristics that can be measured either quantitatively or qualitatively are called *variables*. The term "variable" is ". . . anything that varies or is prone to variation . . ." (*American Heritage Dictionary of the English Language* 1969). For example, the age, sex, and race of an individual are each distinct variables. Whereas a variable is fixed for any particular individual at the moment of observation, it is "prone to variation" as we observe different individuals. The important point is that variables represent characteristics of subjects under study.

A variable consists of a range of possible *values*. These are the "constituents of the characteristics" referred to in question 3. Variable values must be uniquely determined for each individual. That is, if two interviewers approach Mr. Jones and ask his age, they should each record the same value, say 38. If Mr. Jones had a birthday between the two interviews, he would quite correctly have given two different ages. But presumably both interviewers would have obtained the same birth date for Mr. Jones, for example, June 6, 1945.

An alternative explanation might be that the question was posed in a slightly different manner. For example, one interviewer might ask, "Have you ever had high blood pressure?" while another might ask, "Have you ever been told by a doctor that you had high blood pressure?" In this case the interviewers are both measuring the presence of hypertension in a respondent's lifetime but the implied definitions are slightly different.

A third possibility is that the respondent does not understand the question, so he guesses an answer. This leads to what is sometimes referred to as *response error,* which is a special topic requiring special methods that are beyond the scope of this text. We shall assume that the values of a variable are uniquely assigned to each subject or unit of observation.

The example of age brings out a second important characteristic of variables. Specifically, variables may be either categorical or continuous. Similarly, they may be either quantitative or qualitative. Age is usually considered a *continuous* variable. In a practical sense, continuity means that the underlying variable may be measured more and more finely. For example, although age is commonly measured in years, it could be measured in months, days, hours, minutes, and so forth. The specificity that is used to report a continuous variable depends to some extent on the precision of the measuring instrument.

Other variables, such as sex, state of birth, or smoking status, are *categorical.* These are variables that have distinct separations among the possible values. Most individuals can be classified uniquely as male or female, or as having been born in one of the 50 states or born outside the United States, or as former smokers, current smokers, or ex-smokers. As with continuous variables, it is important that each subject be assigned uniquely to a level or value of a categorical variable.

It should be evident that a continuous variable can be condensed into a categorical variable. To condense a variable means to reduce the number of possible values it can have. For example, with age it is possible to define a new variable, called "generation," with values assigned as follows:

Child: $0 \leq \text{age} < 18$
Young adult: $18 \leq \text{age} < 45$
Middle adult: $45 \leq \text{age} < 65$
Older adult: $65 \leq \text{age}$

Clearly, each individual with a measured age can be classified as a child, young adult, middle adult, or older adult. The new variable has four levels, which exhaustively subdivide the population.

It can be argued that condensing continuous variables causes a number of unnecessary difficulties. This occurs in a number of ways. For example, condensing a continuous variable into a few categories may obscure important differences among groups of observations. Second, statistical techniques such as regression analysis and factor analysis are no longer available to the researcher. Third, when data are gathered as categorical variables it is difficult, if not impossible, to reconstruct the underlying continuous measure. Finally, any particular condensation may be quite arbitrary.

On the other hand, the skillful use of condensations may help an investigator understand relationships across the resulting broad categories. For example, although age is continuous, the study of medical care utilization immediately suggests analyzing certain age groups separately. Children tend to use pediatric care, whereas persons over 65 may have special sources of reimbursement. Hence very different patterns of utilization will occur in broad age categories.

Variables can also be classified as either quantitative or qualitative. The levels of *quantitative* variables correspond to a meaningful ordering. Thus a person age 20 is younger than a person age 25. An honor student is academically superior to a failing student. In contrast, we have *qualitative* variables such as sex or place of residence. These usually cannot be ordered in any meaningful way.

The two dimensions by which variables may be classified yield the measurement scales shown in Table 1.3. Nominal variables are both qualitative and categorical. They merely subdivide a population into mutually exclusive groups such as different hospitals or races. Ordinal variables are categorical and quantitative. These variables also subdivide the population, but they arrange the groups into a hierarchy.

Numerical variables assign each member of the population a numeric value that has inherent meaning. Numerical variables include both interval and ratio scales. An *interval scale* is one where the distance between successive values has a constant meaning. For example, a change of 5°F in temperature is the same whether it begins at 0° or 50°. *Ratio scales* have a "true zero point." For example, weight and height are ratio scales.

Table 1.3 Scales of Measurement

	Density	
Character	Categorical	Continuous
Quantitative	Ordinal	Numerical
Qualitative	Nominal	Special

Finally, it is possible that a variable be both qualitative and continuous. An example is the perception of the color spectrum. Similarly, eating leads to a combining of taste, smell, and even appearance. An example that occurs in research is when several continuous variables are measured on subjects. It is generally not possible, without statistical maneuvers, to give these simultaneous measures inherent meaning. Fortunately, variables that are continuous and qualitative are outside the scope of this text.

In what follows, attention will be focused on categorical variables. The principal concern will be on the relationships among several categorical variables. Occasionally, as in logistic regression, the use of continuous variables such as age will be considered. The important point is that the response variable is categorical, but the factors or independent variables may be categorical in general or continuous for binary responses.

1.2 CONTEXT AND OBJECTIVES

Categorical data analysis has been a statistical methodology for as long as correlation analysis (Pearson 1904). However, the development of techniques for analyzing several categorical variables occurred much later than the methodologies for several continuous variables. The latter can be analyzed with a variety of techniques based on the linear model. Analogous techniques for categorical data began to be developed during the 1960s. At that time, a variety of researchers published papers discussing an approach which became known as the log-linear model. Based on the work of Bartlett (1935) and Roy and Kastenbaum (1956), this approach relied on maximum likelihood estimation and addressed issues primarily associated with independence among sets of categorical variables. Much of this work is summarized in books by Bishop et al. (1975), Everitt (1977), Fienberg (1980), Goodman (1978), Haberman (1978, 1979), and Upton (1978).

A parallel approach based on the linear model and weighted least squares was proposed by Grizzle, Starmer, and Koch (1969). The approach, sometimes called GSK, is computationally simpler than likelihood-based procedures. The method also allows easy estimation of complex functions of the categorical response. This permits hypotheses other than those suggested by independence to be examined. The method of estimation was based on the statistics proposed by Wald (1943) and Neyman (1949). The equivalence of the Wald and Neyman statistics was proved by Bhapkar (1966). Much of the work on the GSK approach is summarized in the text by Forthofer and Lehnen (1981).

A third approach is logistic regression, which is appropriate for a binary response variable. This has the particular advantage of permitting consideration of continuous independent variables or factors. The technique reduces to the log-linear model if all the factors are categorical. A discussion of this methodology is found in Cox (1970) and Nerlove and Press (1973).

The preceding texts have focused on one or the other approaches to categorical data analysis. This has suggested a competition among the approaches which is probably both unintended and inappropriate, since noted by Rao (1973, p. 418), ". . . all three statistics are equivalent . . . [since] they have the same a.d. [asymptotic distribution]." Moreover, each procedure has been implemented in software packages such as BMDP, SAS, and SPSS.

A fourth class of procedures for categorical data analysis focuses on survival data. These can be thought of as categorical methods since the response is often a binary outcome such as death. However, since the underlying random variable is the "survival time," survival analysis is more appropriately thought of as a continuous-variable methodology. Survival analysis is discussed in the texts by Cox and Oakes (1984), Gross and Clark (1975), Kalbfleisch and Prentice (1980), or Lee (1980).

This book introduces and develops each of the three techniques: weighted least squares or GSK methods, maximum likelihood or log-linear models, and logistic regression. The basic philosophy is to utilize whichever technique is most appropriate for a given study design. Given equal levels of appropriateness, the computationally most convenient will be employed. When there are particular advantages to a procedure, they will be pointed out. The emphasis throughout is on substantive interpretation of the statistical analysis.

The level of discussion assumes that readers have completed a one-semester course in introductory statistics or biostatistics. Hopefully, the more mathematically sophisticated reader will find the application of the techniques of interest. In addition, it is assumed the reader has had at least one course in research design.

1.3 CHAPTER SUMMARIES

The discussion begins with a review of probability and statistics in Chapter 2. Here the fundamental axioms of probability are presented together with a definition of conditional probability. These are applied by using Bayes' theorem to prove the odds-ratio approximation to the relative risk.* The basic probability distributions used in categorical data analysis are presented in Sections 2.2–2.4. These include the binomial, multinomial, and Poisson distributions. Parameters are then defined together with various tests for goodness of fit (Sections 2.5–2.7). The chapter concludes with an extended analysis of two sets of infant mortality rates (Section 2.8). Both maximum likelihood and weighted least squares estimators are used together with Pearson, Neyman, and likelihood ratio chi-square goodness-of-fit tests.

The classical two-way contingency table is presented in Chapter 3. In Section 3.1 the alternative hypotheses of independence and homogeneity are compared. The 2×2, $I \times 2$, and $I \times J$ tables are discussed in Sections 3.2–3.4. The usual test for trend is illustrated in Section 3.3.2. Hypotheses of quasi-independence (Section 3.4.3) and partition of chi-square (Section 3.4.2) are presented as methods for summarizing $I \times J$ tables. Measures of association, including the odds ratio and relative risk, are discussed in Section 3.5.

Three-dimensional tables are analyzed in Chapter 4. The discus-

*The relative risk is the ratio of two conditional probabilities: namely, the probability of an event in a fixed time period given the presence of a condition to the same conditional probability given the absence of the same condition.

sion begins by considering simple additive models for tests of homogeneity (Section 4.1). The log-linear model is presented in the context of testing independence (Sections 4.3 and 4.4). The all-possible-models approach to model selection is examined next (Section 4.5). The concept of collapsibility is introduced in Section 4.6. The chapter concludes with an extended discussion of the interpretation of the various models (Section 4.7). Graphical aids to the analysis of the models are also developed.

The analysis of multifactor, one-response models is presented in Chapter 5. Both additive and logistic models are considered with a major focus being the problems of interpreting complex interactions (Sections 5.1–5.4). Techniques analogous to path analysis are developed. Fitted values and prediction equations are utilized for interpreting the results.

Complex tables involving four or more categorical variables are introduced in Chapter 6. Most of the discussion relies on the log-linear model and a relatively unstructured analysis (Section 6.1). Model selection procedures and residual analysis are presented in Sections 6.2 and 6.3. The concept of collapsibility with geometric interpretations is discussed in Sections 6.4 and 6.5.

The logistic regression model with continuous variables is presented in Chapter 7. The approach emphasizes the similarity of logistic regression to the more familiar multiple regression techniques (Sections 7.1–7.3). Conditional probability models, which function like path analysis, are presented in Section 7.4. The chapter concludes with a technique for analyzing symmetric tables (Section 7.6).

Exact test procedures are found in Chapter 8. Section 8.2 begins by considering Fisher's test for 2×2 tables. The Mantel–Haenszel test for $2 \times 2 \times K$ tables is discussed in Section 8.3. This chapter relies heavily on the results for the hypergeometric distribution. The book concludes with a brief discussion of a method for the analysis of complex sample survey data (Section 8.4). Here, weighted least squares methods are used exclusively.

In each chapter the issues are developed through the use of examples. Wherever possible, examples are abstracted from published literature. These include both a variety of data sets used by statisticians for several years and data sets recently published in the health, medical, and social science literature.

2

Univariate Analysis

The analysis of categorical data depends on identifying a probability model that gives rise to observed data. This model can be described in terms of the probabilities of specific events occurring. The complete set of probabilities is termed a *distribution*. Often, the objective of statistical analysis is to simplify this model by identifying a small set of parameters that characterize the distribution. We say that the data "fits" the model when the link between our model and the data is "strong." In other situations, the data may provide evidence against our models. This evidence may suggest alternative parameterizations or radically different models.

In this chapter, a variety of simple probability models are considered. The axioms for probability models are described and illustrated in Section 2.1. These are applied in Sections 2.2–2.4 to develop the most commonly used models, including the binomial, multinomial, and Poisson distributions. Analysis of data requires being able to estimate model parameters and to test the fit of data to models, which is done in Sections 2.5 and 2.6. Finally, in Sections 2.7 and 2.8 we consider the more complex problem of assessing the fit of a data set to a particular model.

The reader interested in exploring probability distributions in greater detail should read a basic text in probability theory such as that by Fisz (1963) or Harris (1966). A comprehensive review of discrete probability distributions is given in Johnson and Kotz (1969).

2.1 PROBABILITY

Categorical variables divide populations of objects into distinct groups or subsets. For example, a hospital can be located in only one region of the country, the West or South but not both. The division of a population into groups means that we can think in terms of the proportion of the population in each group or category. That is, what proportion of hospitals fall into the West region? These proportions are an alternative to counts as parameters for summarizing population characteristics.

In practice, we usually select a sample from the population. In this circumstance we ask what is the "probability" of selecting a hospital that is in the western region if all hospitals have a chance of selection that is unaffected by their region. This probability corresponds to the proportion in any given region, and the correspondence depends on the method of selection. For the most part we assume equal probabilities of selection. Generalizing these ideas leads to the "axioms of probability."

Sections 2.1.1–2.1.3 give some general results for probabilities. Sections 2.1.4 and 2.1.5 provide mathematical definitions of expectations and some useful tricks for proving results. These should be referred back to on an "as needed" basis.

2.1.1 Axioms

Suppose that one hospital is selected from among the 8080 short-stay hospitals in the United States in 1981 in such a manner that (1) every hospital is equally likely to be selected, and (2) the selection is at "random." Given these assumptions, the probability of selecting a hospital from a particular region is simply the proportion of all hospitals in that region. The bottom row of Table 2.1 shows the number and proportion of hospitals in each of four regions. The proportion for the Northeast is 0.147. If the selection is random, the probability of the

Table 2.1 Distribution of Short-Stay Hospitals by Geographic Region and Bed Size, United States, 1981

		Region				
Bed Size	Parameter[a]	Northeast (1)	North Central (2)	South (3)	West (4)	Total[b] (5)
6–49	N	223	901	1679	734	3537
	p	0.028	0.112	0.208	0.091	0.438
50–99	N	301	487	751	379	1918
	p	0.037	0.060	0.093	0.047	0.237
100–199	N	298	414	491	224	1427
	p	0.037	0.051	0.061	0.028	0.177
200–299	N	196	168	169	99	632
	p	0.024	0.021	0.021	0.012	0.078
300+	N	167	186	146	67	566
	p	0.021	0.023	0.018	0.008	0.070
Total[b]	N	1185	2156	3236	1503	8080
	p	0.147	0.267	0.400	0.186	1.000

[a] N, Population size; p, cell probability.
[b] *Marginal* counts and proportions.
Source: National Center for Health Statistics (1983).

selected hospital being from the Northeast is 0.147. This may seem to be belaboring the obvious, but the key point is that the concepts of probability and random selection are bound up together. For our purposes, the notion of probability corresponds to underlying, possibly unknown, proportions.

In what follows it will be convenient to use a more general representation. Let the variable A take on several values $i = 1, \ldots, I$. The probability that it takes on a particular value we denote

$$P(A = i) = p_i. \tag{2.1}$$

We assume that $0 \leq p_i \leq 1$ and $(p_1 + p_2 + \cdots + p_I) = \Sigma_{i=1}^{I} p_i = 1$. Using our hospital example, let $i = 1$ indicate Northeast, $i = 2$ indicate

North Central, and so forth for the variable region (R), then $P(R = 1) = p_1 = 0.147$, $P(R = 2) = p_2 = 0.267$, $P(R = 3) = p_3 = 0.400$, and $P(R = 4) = p_4 = 0.186$.

We require one more assumption, that of *additivity*. Suppose that an individual is in one of two levels of a variable. The probability of selecting such an individual is the sum of the two probabilities. For example, what is the probability of selecting a hospital from the North, that is, either from the North Central or from the Northeast? The categorical variable divides the population into mutually exclusive categories, so $R = 1$ or $R = 2$ but not both. The meaning of additivity is that

$$P[(R = 1) \text{ or } (R = 2)] = P(R = 1) + P(R = 2) = p_1 + p_2$$
$$= 0.147 + 0.267 = 0.413.$$

In general terms, when two events are mutually exclusive, the probability of the joint event is the sum of the individual events.

As a further illustration suppose that we wanted the probability of a non-South hospital. This can be obtained in two ways. First is to note that non-South means North or West. Using the previous result, we obtain

$$P(\text{non-South}) = P(\text{North or West}) = P(R = 1 \text{ or } R = 2 \text{ or } R = 4)$$
$$= 0.41 + 0.19 = 0.60$$

(except for rounding error). The alternative is to recognize that $P(\text{South}) + P(\text{non-South}) = 1$, so

$$P(\text{non-South}) = 1 - P(\text{South}) = 1 - 0.40 = 0.60.$$

To sum up, probabilities associated with a categorical variable, denoted A, with I levels indexed by $i = 1, 2, \ldots, I$, have three important properties:

$$(1) \quad P(A = i) = p_i, \text{ where } 0 \le p_i \le 1 \tag{2.2}$$

$$(2) \quad P(A = 1) + P(A = 2) + \cdots + P(A = I) = \sum_{i=1}^{I} p_i = 1 \tag{2.3}$$

$$(3) \quad P(A = i \text{ or } A = j) = p_i + p_j \quad i \neq j. \tag{2.4}$$

When these properties hold, the complete set of probabilities,

$$P(A = 1) = p_1, \quad P(A = 2) = p_2, \quad \ldots, \quad P(A = I) = p_I,$$

is called the *probability distribution*. Notice that for the complement of the event $(A = i)$ we have

$$P(A \neq i) = 1 - P(A = i). \tag{2.5}$$

We will often write this as $P(\overline{A}) = 1 - P(A)$, where A is shorthand for $A = i$ and $\overline{A}$ is shorthand for $A \neq i$.

2.1.2 Conditional Probability and Independence

Usually, we analyze the relation between pairs of variables. For example, the question "What is the probability that a hospital randomly selected from the South has 50–99 beds?" implies that we are interested in the association between bed size and geographic region. More generally, we want to discuss the probability that variable A equals i ($i = 1,2, \ldots, I$) whenever variable B equals j ($j = 1,2, \ldots, J$). This is a *conditional probability*.

Suppose that $P(A = i) > 0$ for all i and $P(B = j) > 0$ for all j. Conditional probabilities are written in terms of the "joint probabilities" and "marginal probabilities":

$$\begin{aligned} P(A = i|B = j) &= \frac{P(A = i, B = j)}{P(B = j)} \\ P(B = j|A = i) &= \frac{P(A = i, B = j)}{P(A = i)}. \end{aligned} \tag{2.6}$$

The former equation is read "The probability that A equals i given B equals j, equals the probability that A equals i and B equals j, divided by the probability that B equals j."

The term $P(A = i, B = j)$ in the numerator of (2.6) is termed the *joint probability* that $A = i$ *and* $B = j$. The denominators of (2.6) are termed *marginal probabilities* because

$$[P(A = 1, B = j) + P(A = 2, B = j) + \cdots + P(A = I, B = j)]$$
$$= \sum_{i=1} P(A = i, B = j) = P(B = j)$$
$$[P(A = i, B = 1) + P(A = i, B = 2) + \cdots + P(A = i, B = J)]$$
$$= \sum_{j=1} P(A = i, B = j) = P(A = i). \tag{2.7}$$

A marginal probability is merely the joint probability of two variables summed over the levels of one of the variables.

Independence is one way in which statisticians specify that two variables are unrelated. Variables are termed *independent* whenever the following holds for all i and j:

$$P[(A = i) \text{ and } (B = j)] = P(A = i, B = j) = P(A = i)P(B = j). \tag{2.8}$$

When (2.8) does not hold, we say that A and B are dependent. That is, the probability that $A = i$ depends on which level of B is considered and conversely. We write this as

$$P(A = i, B = j) = P(B = j)P(A = i|B = j) = P(A = i)P(B = j|A = i)$$
$$i = 1, \cdots, I \quad \text{and} \quad j = 1, \cdots, J. \tag{2.9}$$

Notice that the joint probability is defined symmetrically in terms of the product of the conditional and marginal probabilities. Finally, notice that if A and B are independent, $P(A|B) = P(A)$ and $P(B|A) = P(B)$.

These general ideas can be illustrated with the hospital example. Table 2.1 shows the bed-size distribution of hospitals, a categorical variable with five levels. The table is cross-classified by another categorical variable, region, which has four levels. Thus $B = 1$ corresponds to bed size 6–49, $B = 2$ to 50–99, and so forth. Clearly, if a hospital is selected at random, the probability that it has from 100 to fewer than 200 beds is $1427/8080 = 0.177$.

Do we have enough information to determine the probability of selecting a western hospital with under 50 beds? Yes: $P(B = 1 \text{ and } R = 4) = 734/8080 = 0.091$. A more interesting question might be whether these two variables are independent. The answer to this depends on the definition of independence (2.9). To apply the definition, we compute the product of the marginal probabilities: $P(B = 1)P(R =$

4) = 0.438(0.186) = 0.081. Since this is different from the value obtained from the complete table or "joint distribution" (0.091), it is clear that bed size and region are not independent.

An alternative way of viewing this is to ask whether the row and column totals contain all the "information" in the table or joint distribution. If the variables are independent, they do. Specifically, if bed size and region are independent, the product of the marginal probabilities yields the corresponding joint probability. Consider the largest bed size in the South. If these variables are independent, the cell entry would be $P(B = 5)P(R = 3) = (566/8080)(3236/8080) = 0.070(0.400) =$ 0.028. The actual value is 0.018, which shows that the table contains more information than is found in its margins.

Some of the terminology is now clearer. *Joint distribution* means the complete collection of cell proportions from the body of the table. *Marginal distribution* refers only to the row and column proportions.

Table 2.2 Conditional Distributions of Short-Stay Hospitals Given Bed Size and Region, United States, 1981

Bed Size (B)	Parameter (Region|Bed, Bed|Region)	Region (R): Northeast (1)	North Central (2)	South (3)	West (4)	Total (5)
6–49	$R\|B$	0.063	0.255	0.475	0.208	1.000
	$B\|R$	0.188	0.418	0.519	0.488	0.438
50–99	$R\|B$	0.157	0.254	0.391	0.198	1.000
	$B\|R$	0.254	0.226	0.232	0.252	0.237
100–199	$R\|B$	0.209	0.290	0.344	0.157	1.000
	$B\|R$	0.251	0.192	0.152	0.149	0.177
200–299	$R\|B$	0.310	0.266	0.267	0.157	1.000
	$B\|R$	0.165	0.078	0.052	0.066	0.078
300+	$R\|B$	0.295	0.329	0.258	0.118	1.000
	$B\|R$	0.141	0.086	0.045	0.045	0.070
Total		0.147	0.267	0.400	0.186	—
		1.000	1.000	1.000	1.000	

Finally, the *conditional distributions* can be in terms of either the rows or columns. Table 2.2 displays the bed-size conditional distribution together with the regional conditional distribution. For example, the conditional probability of a smallest-bed-size hospital being from the South is 1679/3537 = 0.475. Similarly, the probability of a western hospital having 300 or more beds is 67/1503 = 0.045.

2.1.3 Bayes' Theorem

In a variety of situations it is useful to be able to derive one set of conditional probabilities from a different set. For example, among high school graduates, what proportion are women? Suppose that we had only the conditional distribution data on education (E) given sex (S) shown in Table 2.3, and we know from other sources that women represent 45 percent of the target population. We have $P(E = i|S = j)$ from Table 2.3 and $P(S = 1) = 0.45$, but what we want to know is $P(S = 1|E = 2)$. Using our definition of conditional probability (2.6), we can write

$$P(S = 1|E = 2) = \frac{P(S = 1, E = 2)}{P(E = 2)} \tag{2.10}$$

$$= \frac{P(S = 1)P(E = 2|S = 1)}{P(S = 1, E = 2) + P(S = 2, E = 2)}.$$

Table 2.3 Conditional Distribution of Education Given Sex ($E|S$)

	Education (E)			
Sex (S)	College Degree (1)	High School Diploma (2)	Neither (3)	Total
Female (1)	0.22	0.44	0.33	1.00[a]
Male (2)	0.18	0.27	0.55	1.00
Total	0.20	0.35	0.45	1.00

[a]Row total differs because of rounding error.

The denominator of (2.10) is obtained from (2.7). Finally, applying (2.9), we have

$$P(S = 1|E = 2) = \frac{P(S = 1)P(E = 2|S = 1)}{P(S = 1)P(E = 2|S = 1) + P(S = 2)P(E = 2|S = 2)}$$

$$= \frac{0.45(0.44)}{0.45(0.44) + (0.55)(0.27)} = 0.57.$$

This is an example of Bayes' theorem. In general,

$$P(A = i|B = j) = \frac{P(A = i)P(B = j|A = i)}{\sum_{i=1}^{I} P(A = i)P(B = j|A = i)} \tag{2.11}$$

Notice that the denominator of (2.11) equals $P(B = j)$.

An important application of Bayes' theorem occurs in the analysis of case-control studies. This study design gathers data on exposure from subjects with a particular disease (cases) and compares the exposure rate to that found in an otherwise comparable group of disease-free subjects (controls). Thus one can compare the conditional probabilities of exposure and assess the relationship between exposure (E) and disease occurrence (D). The underlying probabilities can be arrayed as shown in Table 2.4. Unfortunately, what is desired is the comparison of the conditional probabilities of disease given exposure to the conditional probabilities among the unexposed. The latter comparison can readily be obtained from a cohort study, the structure of

Table 2.4 Probability Structure of a Case Control Study

	Disease Status (D)	
Exposure Status (E)	Present	Absent
Present	$P(E\|D)$	$P(E\|\overline{D})$
Absent	$P(\overline{E}\|D)$	$P(\overline{E}\|\overline{D})$
Total	1.0	1.0

Table 2.5 Probability Structure of a Cohort Study

Exposure Status (E)	Disease Status (D)		
	Present	Absent	Total
Present	$P(D\|E)$	$P(\bar{D}\|E)$	1.0
Absent	$P(D\|\bar{E})$	$P(\bar{D}\|\bar{E})$	1.0

which is shown in Table 2.5. Note that in Tables 2.4 and 2.5 we have indicated the absence of an event by placing a bar over the corresponding letter.

We can link the two study designs by writing the disease risks, or incidence rates in a given time period, as conditional probabilities:

$$P(D|E) = \frac{P(D)P(E|D)}{P(D)P(E|D) + P(\bar{D})P(E|\bar{D})} \tag{2.12}$$

$$P(D|\bar{E}) = \frac{P(D)P(\bar{E}|D)}{P(D)P(\bar{E}|D) + P(\bar{D})P(\bar{E}|\bar{D})} \tag{2.13}$$

where $\bar{E}$ denotes unexposed and $\bar{D}$ denotes no disease. One way to compare the conditional probabilities is to take their ratio. This is termed the *relative risk* by epidemiologists:

$$RR = \frac{P(D|E)}{P(D|\bar{E})}$$

$$= \frac{P(D)P(E|D)/[P(D)P(E|D) + P(\bar{D})P(E|\bar{D})]}{P(D)P(\bar{E}|D)/[P(D)P(\bar{E}|D) + P(\bar{D})P(\bar{E}|\bar{D})]}$$

Now recall that $P(\bar{D}) = 1 - P(D)$, so we obtain

$$RR = \frac{P(E|D)\{P(\bar{E}|\bar{D} + P(D)[P(\bar{E}|D) - P(\bar{E}|\bar{D})]\}}{P(\bar{E}|D)\{P(E|\bar{D} + P(D)[P(E|D) - P(E|\bar{D})]\}}.$$

The final step is to make the "rare disease assumption." This means that we assume $P(D)$ is so small that the terms in brackets on the right side of the equation can be considered zero, yielding

$$RR \approx \frac{P(E|D(P(\overline{E}|\overline{D})}{P(\overline{E}|D)P(E|\overline{D})} = \text{OR} \tag{2.14}$$

This final term is called the odds, or cross-product, ratio. It has the advantage that it can be estimated from Table 2.4. This result was first derived by Cornfield (1951).

2.1.4 Expectation and Variance

In what follows it will be necessary to refer to the average value of a random variable. Suppose that $Y = i$, $i = 0, \ldots, I$ is a random variable; then its average or expected value is given by $E(Y) = \Sigma_{i=0}^{I} iP(Y = i)$. The variance of Y is given by

$$\begin{aligned} \text{Var}\,(Y) &= E[Y - E(Y)]^2 = \sum_{i=0}^{I} [i - E(Y)]^2 P(Y = i) \\ &= E(Y^2) - [E(Y)]^2. \end{aligned} \tag{2.15}$$

Finally, it is sometimes convenient to write the variance in terms of its *factorial moments*:

$$\text{Var}\,(Y) = E[Y(Y-1)] + E(Y) - [E(Y)]^2. \tag{2.16}$$

2.1.5 Some Brief Notes on Combinations (Harris 1966)

Sometimes we want to count the number of ways in which samples of a given size can be obtained. This is the same as the binomial coefficient:

$${}_nC_i = \frac{n!}{i!(n-i)!} \qquad 0 \le i \le n \tag{2.17}$$

where $n! = n(n-1)(n-2) \cdots (2)(1)$. This is read as "$n$ choose i" and is derived in the *binomial theorem*:

$$(a + b)^n = \sum_{i=0}^{n} {}_nC_i a^i b^{(n-i)} \tag{2.18}$$

Three useful equalities may be derived from this theorem:

$$(1) \quad \sum_{i=0}^{\infty} {}_nC_i = 2^n \tag{2.19}$$

$$(2) \quad \sum_{i=0}^{\infty} (-1)^i {}_nC_i = 0 \tag{2.20}$$

$$(3) \quad \sum_{i=0}^{\infty} {}_nC_i \, {}_mC_{j-i} = {}_{n+m}C_j \tag{2.21}$$

Finally, the base of the natural logarithms is:

$$e^t = \sum_{i=0}^{\infty} \frac{t^i}{i!} \qquad \text{where } e = \sum_{i=0}^{\infty} \frac{1}{i!}. \tag{2.22}$$

2.2 THE BINOMIAL DISTRIBUTION

Survey organizations are often interested in evaluating the success rate of new interviewers. One measure of success is the ability to obtain a completed interview. A common technique is to train an interviewer and then have him or her put the training to work in a field trial. In many surveys it is reasonable to expect that an interviewer can obtain 90 percent of the assigned interviews. Suppose that after a training period, the interviewer was assigned 15 eligible adults and was able to obtain 11 interviews. Is this reasonable? The answer requires developing a statistical model.

Model development begins by carefully identifying a set of assumptions. These lead to a probability distribution for evaluating the interviewer's performance. Two assumptions are fairly straightforward:

1. The number of eligible adults is fixed at 15.
2. An eligible adult can either be interviewed or not; that is, the response is binary.

The next two assumptions are more probabilistic:

3. The interviewing of one adult does not affect the chance of interviewing any other adult.

4. The chance of a completed interview is the same for each adult.

That is, assume that the interviews are independent and the interview rate is constant across the sample. In shorthand we say that the sample observations are *independently and identically distributed* (*i.i.d.*). Finally:

5. The sample of eligible adults is drawn from a pool of individuals which is similar in its probability of being interviewed to those usually interviewed by the survey organization.

The next step in building a model is to write the problem in symbols. Let $U_h = 1$ for a completed interview and 0 otherwise, where h indexes the 15 individuals, and let Y denote the total number of completed interviews:

$$\sum_{h=1}^{15} U_h = Y. \tag{2.23}$$

The independent and identical distribution assumptions (3 and 4) can be written in terms of the random variable U_h:

$$\text{Identical distribution:} \quad P(U_h = 1) = p > 0, \qquad h = 1, \ldots, 15 \tag{2.24}$$

$$\text{Independence:} \quad P(U_h = u, U_{h'} = u') = P(U_h = u)P(U_{h'} = u')$$
$$\text{where} \quad u, u' = 0, 1; \quad h, h' = 1, 2, \ldots, 15; \quad h \neq h'. \tag{2.25}$$

Our fifth assumption leads to $p = 0.9$. Note that U_h is binary and implicitly the sample size is fixed at 15 (assumptions 1 and 2).

Now let us consider the distribution of Y, the number of completed interviews. If all 15 adults had been interviewed, then $U_1 = 1, U_2 = 1, \ldots, U_{15} = 1$, and $Y = \Sigma_{h=1}^{15} U_h = 15$. If only 14 adults had been interviewed, one way is $U_1 = 0, U_2 = 1, \ldots, U_{15} = 1$. Another way for $Y = 14$ to occur is if $U_1 = 1, U_2 = 0, U_3 = 1, \ldots, U_{15} = 1$. In fact, there are 15 ways we can get $Y = 14$. The generalization of this is termed *combinations* and written

$${}_nC_y = \frac{n!}{y!(n - Y)!} = \frac{n(n-1)\cdots 1}{y(y-1)\cdots 2(1)(n-y)(n-y-1)\cdots 1}. \tag{2.26}$$

Using the i.i.d. assumptions, we compute the probability of obtaining 15 interviews: $P(Y = 15) = P(U_1 = 1, U_2 = 1, \ldots, U_{15} = 1) = P(U_1 = 1)P(U_2 = 1) \cdots P(U_{15} = 1) = (0.9)(0.9) \cdots (0.9) = (0.9)^{15} = 0.2059$. Thus there is a 20.6 percent chance of obtaining 15 interviews. What about 14 interviews?

$$\begin{aligned} P(Y = 14) &= {}_{15}C_{14}P(U_1 = 1, U_2 = 1, \ldots, U_{15} = 0) \\ &= \frac{15!}{1!14!} P(U_1 = 1)P(U_2 = 1)P(U_3 = 1) \cdots P(U_{15} = 0) \\ &= 15(0.9)^{14}(0.1) = 0.3432. \end{aligned}$$

Note that $P(U_h = 0) = 1 - P(U_h = 1) = 0.1$. Continuing for 13 and 12 interviews yields

$$P(Y = 13) = {}_{15}C_{13}(0.9)^{13}(0.1)^2 = 105(0.2542)(0.01) = 0.2669,$$

$$P(Y = 12) = {}_{15}C_{12}(0.9)^{12}(0.1)^3 = 455(0.2824)(0.001) = 0.1285.$$

Now we can ask how unusual our trainee is:

$$\begin{aligned} P(Y \le 11) &= 1 - P(Y > 11) \\ &= 1 - [P(Y = 12) + P(Y = 13) + P(Y = 14) + P(Y = 15)] \\ &= 1 - (0.1285 + 0.2669 + 0.3432 + 0.2059) \\ &= 1 - 0.9445 = 0.0555. \end{aligned}$$

Thus our trainee is just above the fifth percentile of interviewers, which may not be acceptable.

The general form of the binomial is as follows. Assume that we have n trials each of which leads to a success or failure. Assume that the probability of a success, p, is fixed for each of the n trials and that the trials are independent. Let Y denote the number of successes. Then Y has a binomial distribution and

$$P(Y = y) = {}_nC_y p^y(1 - p)^{n-y} \qquad y = 0, 1, \ldots, n. \tag{2.27}$$

In general, we will let lowercase letters denote observed values of random variables.

We can derive some properties of the binomial using simple algebra. First, the sum of the binomial probabilities is 1. This we obtain from the binomial expansion of $(p + q)^n$, where $q = 1 - p$: $1 = (p + q)^n = \Sigma_{y=0}^{n} {}_nC_y p^y q^{n-y}$. The expected value of Y is given by

$$E(Y) = \sum_{y=0}^{n} y_n C_y p^y q^{n-y} = np. \tag{2.28}$$

Here $E(Y)$ is read the "expectation of Y" or the "expected number of successes." The proof of (2.28) is straightforward:

$$E(Y) = \sum_{y=1}^{n} y_n C_y p^y q^{n-y} = np \sum_{y=1}^{n} \frac{(n - 1)!}{(y - 1)!(n - y)!} p^{y-1} q^{n-y}.$$

Now let $x = y - 1$ and let $m = n - 1$:

$$E(Y) = np \sum_{x=0}^{n-1} \frac{(n - 1)!}{x![n - (x + 1)]!} p^x q^{n-(x+1)}$$

$$= np \sum_{x=0}^{m} \frac{m!}{x!(m - x)!} p^x q^{m-x} = np.$$

The last step follows because $\Sigma_{x=0}^{m} m! p^x q^{n-x} / x!(m - x)! = (p + q)^m = 1$.

The third important property of the binomial distribution is its variance: Var $(Y) = E(Y^2) - [E(y)]^2$. What we need to know is $E(Y^2)$. Using the approach for the proof of $E(Y) = np$, we can show that $E[Y(Y - 1)] = \Sigma_{y=0} y(y - 1)_n C_y\, p^y (1 - p)^{n-y} = p^2$. Hence $E(Y^2) = n(n - 1)p^2 + np$. This leads immediately to

$$\text{Var}\ (Y) = (n(n - 1)p^2 + np) - (np)^2 = (n^2p^2) - np^2 + np - (np)^2$$

$$= np(1 - p). \tag{2.29}$$

It is extremely tedious to compute the values of $P(Y = y)$ when n is large. But for random variables with a finite mean and variance such as the binomial, an approximation to the normal distribution can be used.

This is sometimes called the *z-score* or *z-transformation*. It has the general form

$$Z = \frac{Y - E(Y)}{\sqrt{\mathrm{Var}\,(Y)}} = \frac{Y - np}{\sqrt{np(1 - p)}}. \tag{2.30}$$

Since Y is categorical and Z is continuous, a *continuity correction* is made by adding or subtracting 0.5 to or from Y.

To see how the approximation and continuity correction are used, consider our earlier example of finding $P(Y \leq 11)$. The "$\leq$" means that we want to find probabilities for all values that are at least 11 but less than 12. The midpoint is 11.5. Thus Y is corrected for continuity by adding 0.5 to 11. The resulting z transformation is

$$z = \frac{y + 0.5 - np}{\sqrt{npq}} = \frac{11 + 0.5 - 15(0.9)}{\sqrt{15(0.9)(0.1)}} = \frac{11.5 - 13.5}{\sqrt{1.162}} = -1.72.$$

Using a standard normal table, we have $P(Y \leq 11) = P(Y < 11.5) = P(Z \leq -1.72) = 0.043$. Notice this is a little less than we obtained directly. Generally, the approximation works well when both $np > 5$ and $nq > 5$. In the present example, $nq = 15(0.1) = 1.5$, so the discrepancy is to be expected.

Suppose that we had a probability of success of 0.4 and $n = 20$. By exact computation we can see that $P(Y \geq 10) = 0.245$. What is the corresponding normal approximation? First compute the mean and variance: $E(Y) = np = 20(0.4) = 8$, $\mathrm{Var}\,(Y) = npq = 20(0.4)(0.6) = 4.8$. Next apply the continuity correction. This means that we want values of Y greater than 9.5, which is midway between 9 and 10. The z-transformation is $z = (9.5 - 8)/\sqrt{4.8} = 0.68$. Again using a standard normal table, we have $P(Y \geq 10) = P(Y > 9.5) = P(Z > 0.68) = 0.2466$. Here with larger values of $np = 8$ and $n(1 - p) = 12$, the approximation is much closer to the exact value computed earlier.

To summarize, the binomial distribution is appropriate when we have:

1. Binary outcome
2. Fixed sample size, n
3. Identical probability of an outcome, p
4. Independent observations of the outcome

(2.31)

The probability distribution of a binomial random variable Y is given by

$$P(Y = y) = {}_nC_y p^y(1 - p)^{n-y} \quad y = 0, 1, 2, \ldots, n. \qquad (2.32)$$

The mean and variance of a binomial random variable are

$$E(Y) = np \qquad \text{Var}(Y) = np(1 - p). \qquad (2.33)$$

Finally, the Z-transformation is given by

$$Z = \frac{Y_c - np}{\sqrt{np(1 - p)}}. \qquad (2.34)$$

where Y_c indicates the continuity correction:

$$y - 0.5 \quad \text{for } P(Y < y) \text{ or } P(Y \leq y)$$

$$y + 0.5 \quad \text{for } P(Y > y) \text{ or } P(Y \geq y).$$

2.3 THE MULTINOMIAL DISTRIBUTION

Suppose that we modify assumption 1 of the binomial (2.31). For example, interviewers fail to obtain interviews for a variety of reasons. An eligible adult may not be home or may refuse the interview. Now we have three possible outcomes: (1) interview obtained; (2) interview refused; (3) respondent not home. We can represent these possibilities with three indicator variables for the hth respondent:

$$U_{h1} = 1 \text{ if interview obtained,} \quad 0 \text{ otherwise;}$$

$$U_{h2} = 1 \text{ if interview refused,} \quad 0 \text{ otherwise;} \qquad (2.35)$$

$$U_{h3} = 1 \text{ if respondent not home,} \quad 0 \text{ otherwise;} \quad h = 1, 2, \ldots, n \text{ trials.}$$

Notice that only one of the U_{hi} can be 1, so $\Sigma_i U_{hi} = 1$. In addition, even if 11 interviews were obtained, there are now many ways the noninterviews can occur. What interests us is the distribution of the outcomes across the n trials.

The outcomes can be examined in terms of the number of trials falling into each of the categories. That is,

$$Y_i = \sum_{h=1}^{n} U_{hi}. \tag{2.36}$$

This implies that $n = \Sigma_{i=1}^{I} Y_i$ since $\Sigma_{i=1}^{I} Y_i = \Sigma_{i=1}^{I}\Sigma_{h=1}^{n} U_{hi} = \Sigma_{h=1}^{n}\Sigma_{i=1}^{I} U_{hi} = \Sigma_{h=1}^{n} 1 = n$. Let $P(U_{hi} = 1) = p_i$, so $\Sigma_{i=1}^{I} p_i = 1$. This leads to the multinomial distribution:

$$P(Y_1 = y_1, Y_2 = y_2, \ldots, Y_I = y_I) = \frac{n!}{y_1!\, y_2! \cdots y_I!} p_1^{y_1} p_2^{y_2} \cdots p_I^{y_I},$$

or more compactly,

$$P(Y_i = y_i) = \frac{n!}{y_i!} \prod_{i=1}^{I} p_i^{y_i}. \tag{2.37}$$

To continue our example, suppose it is known that refusals and "not at homes" are equally likely. This means that $p_1 = 0.9$, $p_2 = 0.05$, and $p_3 = 0.05$. Suppose that our trainee received three refusals and one not at home. What is the probability of this?

$$P(Y_1 = 11, Y_2 = 3, Y_3 = 1) = 15!\,\frac{(0.9)^{11}}{11!}\,\frac{(0.05)^3}{3!}\,\frac{(0.05)^1}{1!} = 0.0107.$$

Other possible outcomes for the noninterviews are shown in Table 2.6. Notice that the sum over the noninterview categories yields the same result (to within rounding error) as if we had computed the probability of exactly 11 interviews where a binomial applied: $P(Y_1 = 11) = {}_{15}C_{11}(0.9)^{11}(0.1)^4 = 0.0428$. This illustates the idea that the multinomial is a straightforward generalization of the binomial. The question of whether the particular pattern of interview outcomes is unusual is somewhat more difficult to formulate.

The discussion of the example in the context of the binomial suggests that unusually few interviews were obtained. On the other hand, was the "not at home" pattern unusual? That is, what is the proba-

Table 2.6 Multinomial Probabilities When $n = 15$, $p_1 = 0.9$, $p_2 = 0.05$, $p_3 = 0.05$, and $Y_1 = 11$

Outcome (Y)			
Interview Y_1	Refused Y_2, Y_3	Not Home	Probability
11	4	0	0.0027
	3	1	0.0107
	2	2	0.0161
	1	3	0.0107
	0	4	0.0027
11	+	+	0.0429

bility of 0 or 1 not at homes given 11 of 15 interviews? This can be obtained using the law of additivity and conditional probabilities:

$$P[(Y_2 = 3, Y_3 = 1 | Y_1 = 11) \text{ or } (Y_2 = 4, Y_3 = 0 | Y_1 = 11)]$$
$$= P(Y_2 = 3, Y_3 = 1 | Y_1 = 11) + P(Y_2 = 4, Y_3 = 0 | Y_1 = 11).$$

Next apply the definition of conditional probabilities:

$$P(Y_2 = 3, Y_3 = 1 | Y_1 = 11) = \frac{P(Y_1 = 11, Y_2 = 3, Y_1 = 1)}{P(Y_1 = 11)}$$
$$= \frac{(15!/11!3!1!)(0.9)^{11}(0.05)^3(0.05)^1}{(15!/11!4!)(0.9)^{11}(0.1)^4}$$
$$= \frac{(4!/3!1!)(0.05)^3(0.05)^1}{(0.10)^4} = 0.25.$$

Similarly, $P(Y_2 = 4, Y_3 = 0 | Y_1 = 11) = 0.0625$. So we have little evidence that the noninterview pattern was unusual. A more important result is that the conditional distribution for a multinomial is also multinomial.

Finally, let us find the mean and variance of the multinomial. The marginal distribution of the multinomial is binomial, so we immediately obtain

$$E(Y_i) = np_i \qquad \mathrm{Var}\,(Y_i) = np_i(1 - p_i). \tag{2.38}$$

But what about the covariance of two multinomial random variables? The definition of covariance is

$$\mathrm{Cov}\,(Y_i, Y_j) = E(Y_iY_j) - E(Y_i)E(Y_j). \tag{2.39}$$

What we need is $E(Y_iY_j)$. From (2.36) we can write

$$E(Y_iY_j) = E\left[\left(\sum_{g=1}^{n} U_{gi}\right)\left(\sum_{h=1}^{n} U_{hj}\right)\right] = E\left(\sum_{g=1}^{n}\sum_{h=1}^{n} U_{gi}U_{hj}\right) \tag{2.40}$$

$$= \sum_{g=1}^{n}\sum_{h=1}^{n} E(U_{gi}U_{hj}) = n(n - 1)p_ip_j.$$

This follows because of independence across subjects and the fact that each individual is in one and only one category. Using (2.40) and (2.38) in (2.39), we obtain

$$\mathrm{Cov}\,(Y_i, Y_j) = n(n - 1)p_ip_j - (np_i)(np_j) = -np_ip_j. \tag{2.41}$$

2.4 THE POISSON DISTRIBUTION

The multinomial extends the binomial by considering several possible outcomes. But as with the binomial, it requires a fixed sample size, n. Sometimes n can be considered infinite; other times we may not want to treat n as fixed. For example, the chance of an accident occurring in a specified period may be very small, but after many periods, many accidents will have occurred. Here the number of periods is very large. This can be modeled with the Poisson distribution.

Poisson (1837) examined what occurred to the binomial distribution when the sample size got large and the probability of a success on any given trial got small. The results led to the distribution that bears his name. As usual, we need some specific assumptions. The first two are familiar:

1. The trials are conducted independently.
2. The probabilities are constant from trial to trial.

We need two new assumptions:

3. The probability of more than one success in a given trial is small.

A fourth assumption yields the Poisson distribution. Let p denote the chance of one or more successes in a given trial and n denote the number of trials. The Poisson assumption is:

4. As $n \rightarrow \infty$ and $p \rightarrow 0$, $np = r$.

This last term, r, is the parameter of the Poisson distribution.

The preceding assumptions yield the following result:

$$\lim[{}_nC_y p^y (1-p)^{n-y}] = \frac{r^y e^{-r}}{y!} \qquad n \rightarrow \infty, \quad p \rightarrow 0, \quad np = r. \qquad (2\ 42)$$

Here $e = 2.71828\ldots$ is the base of the natural logarithms as defined in (2.22).

Alternatively, Y is a Poisson random variable if

$$P(Y = y) = \frac{r^y e^{-r}}{y!} \qquad y = 0, 1, 2, \ldots . \qquad (2\ 43)$$

This somewhat elaborate formulation can be made more appealing with an example.

Arbous and Kerrich (1951) discuss the analysis of accident statistics. In an example they present data on accidents among engine shunters aged 31 to 35. The data for 155 person–years of exposure are shown in column (6) of Table 2.7. It is reasonable to assume that the number of hours worked in given year is quite large and the chance of an accident is quite small. A somewhat more questionable assumption is that the exposures occur independently across subjects and time periods. Assume that the chance of accident in any given year is 0.5, which means that $r = 0.5$.

The number of accidents for any individual in the course of a year can be from 0 to 365 or more. In fact, only one person had as many as three. The expected number of accidents can be obtained by constructing the probability distribution and applying it to the 155 years of exposure. The first entry in column (4) is $P(Y = 0) = (0.5)^0 e^{-0.5}/0! = e^{-0.5} = 0.606$. Thus there is a 60.6 percent chance of an individual

Table 2.7 Observed and Expected Accidents Among 155 Engine Shunters Aged 31–35, $r = 0.5$

Number of Accidents (1)	$\frac{r}{x}$ (2)	$\frac{r^2}{x!}$ (3)	$P(X = x)$ (4)	Expected (5)	Observed (6)
0	—	1	0.606	93.9	80
1	0.500	0.500	0.303	47.0	61
2	0.250	0.125	0.076	11.8	13
3+	0.167	0.021	0.015	2.2	1
Total years of exposure				155	155

Source: Arbous and Kerrich (1951). With permission from the Biometric Society.

having no accidents in one year. The expected number out of 155 years is 155(0.606) = 93.9, compared to the 80 observed. Following the same reasoning, the probability of exactly one accident is $P(Y = 1) = (0.5)^1 e^{-0.5}/1! = 0.303$, which leads to 155(0.303) = 47.0. That is, we expect 47 people of the 155 to have exactly one accident in any given year. In a similar manner we obtain an expected 11.8 individuals with two accidents and 2.3 with three or more accidents.

The observed and expected numbers of accidents are quite different. With too few at 0 and too many at 1, there is either evidence against the Poisson model being appropriate or against there being a 50 percent chance of an accident in a given year. To explore this, we compute the mean number of accidents expected and observed. The mean number expected is given by $[0(93.9) + 1(47.0) + 2(11.8) + 3(2.3)]/155 = 0.50 = r$. The mean number observed is given by $[0(80) + 1(61) + 2(13) + 3(1)]/155 = 0.581$. So it appears that the chance of an accident is closer to 60 percent than 50 percent.

Notice that the mean of the expected number of accidents is equal to the hypothesized value of r. It happens that this is always true for the Poisson distribution. We shall prove this after showing that the distribution sums to unity. Consider (2.43): $\Sigma_{y=0}^{\infty} r^y e^{-r}/y! = e^{-r}(\Sigma_{y=0}^{\infty} r^y/y!) =$

$e^{-r}e^{r} = 1$. Using the same maneuver as in the binomial, it is easy to show that the mean of the Poisson is r:

$$E(Y) = \frac{\sum_{y=0}^{\infty} y r^{y} e^{-r}}{y!} = \frac{r e^{-r} \sum_{y=1}^{\infty} r^{y-1}}{(y-1)!} = r.$$

The variance of the Poisson is also r. This can be obtained in a manner similar to that used for the variance of the binomial. Specifically,

$$E[Y(Y-1)] = \frac{\sum_{y=0}^{\infty} y(y-1) r^{y} e^{-r}}{y!} = \frac{r^{2} e^{-r} \sum_{y=2}^{\infty} r^{y-2}}{(y-2)!} = r^{2}.$$

Hence $E(Y^2) = r^2 + r$, so $\text{Var}(Y) = (r^2 + r) - (r)^2 = r$. The equality of the mean and variance can be illustrated with our example:

$$\frac{(0-0.5)^2(93.9) + (1-0.5)^2(4.7) + (2-0.5)^2(11.8) + (2-0.5)^2(2.3)}{155} = 0.49,$$

which differs from r only because of rounding error.

The interesting result that the mean and variance of the Poisson are equal can be used to test whether observed data were sampled from a Poisson distribution. We can illustrate the procedure with our example. The mean of the observed values has already been computed, $y = 0.58$. Now compute the usual sample variance,

$$s^2 = \frac{[(0)^2(80) + (1)^2(61) + (2)^2(13) + (3)^2(1)] - 155(0.58)^2}{154}$$

$$= \frac{122 - 52.258}{154} = 0.453.$$

The test statistic is $X^2 = (n-1)s^2/\bar{y} = 154(0.453)/0.581 = 120.1$. This can be compared to a chi-square distribution with $\text{df} = n - 1 = 154$. This corresponds to an unusually small value of a chi-square random variable, so we reject the specific hypothesis of the Poisson model. It appears that our data have a variance that is a bit small to have been generated by a Poisson distribution. This suggests that an alternative distribution should be considered. For an extended discussion of these issues, the reader should read the article by Arbous and Kerrich (1951).

2.5 ESTIMATION

The previous examples focused on deriving probabilistic models for various phenomena. This led to probability distributions which could be used to compute the chance of a particular event occurring. In each case the probability distribution could be completely specified. For example, the interviewer trainee has a 0.9 chance of completing an interview. In the accident example the chance of an accident was initially thought to be 0.5. These quantities are termed the *parameters* of the distribution. This is common usage in that once a distribution is assumed, then specifying particular values for parameters determines specific probabilities that should occur. The parameters of the distributions we have considered are shown in Table 2.8. Unfortunately, in many circumstances the specific value of the parameters cannot be specified in advance, so they must be estimated from available data.

There are two approaches to this problem. The more obvious is simply to use the data to "estimate" the model parameters. Here "estimate" means that the data are summarized in such a way as to produce a reasonable guess as to the parameter value. The second approach is not only to estimate the model parameters but to combine

Table 2.8 Parameters of Some Discrete Probability Distributions for Samples of Size n and the Range of the Random Variable

Distribution	Equation	Parameter	Interpretation	Range
Binomial	(2.32)	p	Chance of success	$y = 0, 1, \ldots, n$
Multinomial	(2.37)	p_i	Chance of category i	$y_i = 0, \ldots, n$ $\Sigma_i y_i = n$
Poisson	(2.43)	r	Average number of events in n exposures	$y = 0, 1, 2, \ldots$

this with an estimate of the precision of the estimate. The former is termed *point estimation* and the latter *interval estimation*.

Let us examine some possibilities. In the interviewer trainee example assume that a binomial distribution is appropriate for the 15 attempted interviews. The trainee obtained 11. A reasonable point estimate for the interviewer's probability of an interview is $\hat{p} = 11/15 = 0.733$. In fact, a general choice for the estimator of p in the binomial is $\hat{p} = Y/n$. Is this a "reasonable guess" at $\hat{p}$? Yes, because $E(p) = E(Y/n) = E(Y)/n = np/n = p$. Since the expected value of the estimator is the parameter being estimated, the estimator is called *unbiased*. This generally useful property is one criterion for a reasonable point estimate.

A second criterion is the variability of the estimator. This depends on the variance of the sampled population. For the binomial the variance of Y is $np(1 - p)$. The variance of our estimator of p is $p(1 - p)/n$ since $\text{Var}\ (\hat{p}) = \text{Var}\ (Y/n) = \text{Var}\ (Y)/n = np(1 - p)/n_2 = p(1 - p)/n$. This is a special case of a general result in introductory statistics. The sample mean is an unbiased estimator for the expected value of the corresponding distribution. It will also have a variance equal to the variance of the distribution divided by the sample size.

Applying this to the multinomial, we have that $\hat{p}_i = Y_i/n$ is unbiased and has variance equal to $p_i(1 - p_i)/n$. So in the interviewer trainee example the estimated refusal rate is $\hat{p}_2 = 3/15 = 0.20$, with variance under the null hypothesis of $p_2(1 - p_2)/n = 0.05(0.95)/15 = 0.00317$. We used the assumed value of $p = 0.05$ to compute the variance of $\hat{p}_2$. Alternatively, we can estimate the variance as well as the parameter by using a variance estimator. In this case it would be natural to replace p_2 by its estimator: $V(\hat{p}_2) = \hat{p}_2(1 - \hat{p}_2)/n = 0.2(0.8)/15 = 0.011$. Notice that we write the estimator of $\text{Var}\ (\hat{p}_2)$ as $V(\hat{p}_2)$. An immediate question is whether $\hat{V}(p_2)$ is unbiased. Unfortunately, it is not. To see this, we compute

$$E\left[\frac{\hat{p}_2(1 - \hat{p}_2)}{n}\right] = \frac{\hat{p}_2 - E(\hat{p}_2{}^2)}{n} = \frac{\hat{p}_2(1 - \hat{p}_2)(n - 1)}{n_2}$$

$$= \frac{\text{Var}\ (\hat{p}_2)(n - 1)}{n}.$$

The bias of $V(\hat{p}_2)$ is therefore

$$B[V(\hat{p}_2)] = \frac{\text{Var}(\hat{p}_2)(n-1)}{n} - \text{Var}(\hat{p}_2) = \frac{-\text{Var}(\hat{p}_2)}{n}.$$

As we can see, the bias of $V(p)$ tends to zero quite rapidly as n becomes large. This property of an estimator being nearly unbiased for large samples is sometimes termed *consistency*.

As a final example consider the estimator of r, the parameter for a Poisson-distributed random variable Y. Suppose that we have n independent observations denoted Y_i. An unbiased estimator of r is $\overline{Y} = \Sigma_{i=1}^{n} Y_i/n$, the sample mean. Again it is unbiased: $E(\overline{Y}) = E(\Sigma_{i=1}^{n} Y_i/n) = \Sigma_{i=1}^{n} E(Y_i)/n = nr/n = r$. The variance of Y is r/n, which can be estimated by $V(\hat{r}) = \overline{Y}/n$. So in the accident example an estimate of r is 0.581 with an estimated variance of $0.581/155 = 0.00375$.

The preceding focused on estimating the particular value of a parameter. An alternative approach is to estimate the values of the parameter in which the analyst is confident. That is, one can construct an interval estimate or confidence interval. This type of estimate provides information about the parameter and about the precision of our estimate at the same time. The interval estimate combines three ideas: (1) an estimator, (2) a standard error, and (3) the distribution of the estimator. The usual form of the $(1 - \alpha)100$ percent confidence interval is

$$\hat{\theta} \pm z_{\alpha/2} \sqrt{V(\hat{\theta})}$$

where $\hat{\theta}$ is the estimator, $z_{\alpha/2}$ is the $(\alpha/2)100$ percentile of the standard normal distribution, and $V(\hat{\theta})$ is the estimated variance of the estimator. For most estimators used in categorical data analysis, it is appropriate to assume that the distribution is normal because of large sample sizes. The 95 percent confidence interval for the engine shunters' accident rate would be $0.581 \pm 1.96 \sqrt{0.00375} = (0.461, 0.701)$. This interval indicates we are 95 percent confident that the true accident rate is between 0.461 and 0.701, assuming that accidents follow a Poisson distribution.

The construction of the confidence interval for the Poisson parameter immediately suggests two refinements. The easiest is to avoid the

Poisson assumption, especially since the earlier test of the Poisson hypothesis was rejected (end of Section 2.4). This alternative confidence interval is constructed using the observed sampling variance, 0.453:

$$0.581 \pm 1.96\sqrt{\frac{0.453}{155}} = (0.475, 0.687).$$

The second refinement is to improve the normal distribution assumption for the estimator. Often, discrete distributions are asymmetric. This suggests that improved confidence intervals can be obtained by working on the logarithmic scale. It can be shown that a good approximation to the variance of 2 log $(\hat{\theta})$ is

$$\text{Var}\,[\log(\hat{\theta})] = \frac{\text{Var}\,(\hat{\theta})}{(\hat{\theta})^2}. \tag{2.44}$$

The $(1 - \alpha)100$ percent confidence interval is then given by

$$\exp\left[\log(\theta) \pm z_{\alpha/2}\sqrt{\frac{V(\hat{\theta})}{\hat{\theta}^2}}\right] \tag{2.45}$$

The asymmetric 95 percent confidence interval for the accident rate under the Poisson model is given by

$$\exp\left[\log(\hat{r}) \pm 1.96\sqrt{\frac{1}{n\hat{r}}}\right],$$

or,

$$\exp\left[\log(0.581) \pm 1.96\sqrt{\frac{1}{155(0.581)}}\right] = \exp(-0.543 \pm 0.207)$$
$$= (0.473, 0.714).$$

If we do not believe the Poisson model, the result is

$$\exp\left[\log(0.581) \pm 1.96\sqrt{\frac{0.453}{155(0.581)^2}}\right] = \exp(-0.543 \pm 0.182)$$
$$= (0.484, 0.698).$$

These four confidence intervals are quite similar because of the large sample size. But in general we would probably use the last one because it does not require the Poisson model and reflects the probable asymmetry of the underlying discrete distribution.

2.6 GOODNESS-OF-FIT TESTS

Assumptions about the process that generated data lead to derivations of probability models. In some cases the parameters of the associated distribution can also be explicitly specified. More commonly, the parameters must be estimated using sample data. This leads to the question of whether the model implied by our assumptions can be tested. That is, if a probability distribution is hypothesized, whether the parameters are estimated or specified, can the data be used to test this hypothesis? The answer is yes and the test statistics are often referred to as *goodness-of-fit tests*.

The basic idea is to compare observed counts to counts that would be expected if the model were correct. If the model specifies the distribution parameters, the expected counts are also implicitly specified. On the other hand, when the parameters are estimated, the expected counts are also estimated. For example, in the engine shunter example (Table 2.7), since $r = 0.5$ is specified the expected counts are also specified [column (5)]. On the other hand, when the parameter is estimated as in Table 2.9, the expected counts must also be treated as estimates [column (5)].

There are three test statistics for comparing the observed and expected counts. The best known of the goodness-of-fit tests is the *Pearson chi-square* test, which has the form

$$X^2 = \frac{\sum_{i=1}^{I} (Y_i - m_i)^2}{m_i} \tag{2.46}$$

where Y_i is the observed count and $m_i = E(Y_i|H_0)$. That is, m_i is the expected count in the ith category when the hypothesized distribution is true (H_0), $i = 1, \ldots, I$. Two other test statistics are in common use for assessing the fit of data-specified distributions. These are the *likeli-*

hood ratio chi-square, G^2, and *Neyman*, or *weighted least squares, chi-square*, Q:

$$G^2 = 2 \sum_{i=1}^{I} Y \ln \left(\frac{Y_i}{m_i}\right) \tag{2.47}$$

$$Q = \sum_{i=1}^{I} \frac{(Y_i - m_i)^2}{Y_i}. \tag{2.48}$$

Notice that X^2 and Q are based on the difference between observed and expected counts, while G^2 is based on the corresponding ratio. Also, whenever an observed count is zero, Q is undefined unless the zero is replaced by a small positive quantity such as 0.5.

Suppose that we want to test whether the engine shunter data in Table 2.7 were obtained from a Poisson distribution with mean $r = 0.5$. The expected and observed counts are given in columns (5) and (6) of that table. Using these values, we obtain

$$X^2 = \frac{(80 - 93.9)^2}{93.9} + \frac{(61 - 47)^2}{47} + \frac{(13 - 11.8)^2}{11.8} + \frac{(1 - 2.2)^2}{2.2}$$

$$= 2.06 + 4.17 + 0.12 + 0.65 = 7.00.$$

The value of X^2 will be large whenever there are large deviations between what is observed and expected under the null hypothesis, but when is the value so large as to indicate that the null hypothesis is false?

It has been shown that this test statistic will have a probability distribution known as the chi-square distribution. The chi-square distribution is actually a family of distributions with a parameter known as *degrees of freedom*. The degrees of freedom are obtained by counting the number of categories and then subtracting the number of estimated parameters and constraints required to compute X^2. In the present case there are four categories. We require that $\Sigma_i m_i = \Sigma_i Y_i = 155$, so there is one constraint. The value of r was specified, so estimates are not required. Hence df $= I - 1 - 0 = 4 - 1 = 3$. Using a table of the chi-square distribution we find that $P(X^2 \geq 6.25) = 0.10$, $P(X^2 \geq$

Table 2.9 Observed and Estimated Expected Accidents Among 155 Engine Shunters Aged 31–35, Estimated $r = 0.581$[a]

Number of Accidents (1)	$\frac{r}{x}$ (2)	$\frac{r^x}{x!}$ (3)	$P(X = x)$ (4)	m_i (5)	y_i (6)	$y_i - m_i$ (7)	$\frac{(y - m)^2}{m}$ (8)
0	—	1	0.559	86.6	80	−6.6	0.50
1	0.581	0.581	0.325	50.4	61	10.6	2.23
2	0.291	0.169	0.094	14.6	13	−1.6	0.18
3+	0.194	0.033	0.022	3.4	1	−2.4	1.69
Total	—	—	1.00	155	155	0	4.60

[a]*m*, expected; *y*, observed.

7.81) = 0.05. So we reject the hypothesis at $\alpha = 0.10$ but accept it at $\alpha = 0.05$. Using this test we have some evidence against the underlying distribution being Poisson with a mean of 0.5 accident per year. The hypothesis completely specified the underlying distribution, but its rejection suggests either that the Poisson model does not apply or $r \neq 0.50$. A narrower hypothesis would leave r unspecified and use an appropriate estimate, $\bar{y} = 0.581$. Using this estimate a new set of expected counts can be estimated. These are shown in Table 2.9, which is constructed in the same manner as Table 2.7. The Pearson X^2

Table 2.10 Comparison of Pearson (X^2), Neyman (Q), and Likelihood Ratio (G^2) Chi-Square Test Statistics

Category	Observed y	Expected m	$\frac{(y - m)^2}{m}$	$\frac{(y - m)^2}{y}$	$2\left[y \ln\left(\frac{y}{m}\right)\right]$
1	80	86.6	0.50	0.54	−12.68
2	61	50.4	2.23	1.84	23.29
3	13	14.6	0.18	0.20	−3.02
4	1	3.4	1.69	5.76	−2.45
Total	155	155.0	4.60	8.34	5.14

is the total for column (8). It is smaller but the degrees of freedom are reduced, df = 4 − 1 − 1 = 2, because *r* is estimated from the data. Using a chi-square table the statistic again falls between the 90th and 95th percentiles, so there is some evidence against the Poisson distribution even if we estimate the parameter. A comparison of the three computations is shown in Table 2.10. Notice that the totals are somewhat different, but it can be shown that as the sample size increases the three tend toward the same chi-square distribution (Rao 1973). There is no mathematical evidence favoring one statistic over the others. We will tend to use any one of these statistics based primarily on computational convenience.

2.7 SAMPLE SIZE AND ZEROS

The Neyman chi-square cannot be computed for data sets with observed zeros. This is because its formula contains observed values in the denominator. A number of authors, Grizzle et al. (1969) and Forthofer and Lehnen (1981) among others, have advocated simply replacing the observed zeros by some small quantity such as 0.5 or $1/aI$, where I is the number of categories being considered. This problem is peculiar to the weighted least squares procedures. However, the convergence of the maximum likelihood procedures may also be enhanced by adjusting the cell counts by a small positive quantity. The implications of such adjustments are not understood at present, although it is evident that any adjustment to the data should be undertaken with care. When sampling zeros occur, we will follow the practice of adding 0.5 to the affected cells only when a weighted least squares analysis is undertaken. A more extended discussion of alternative adjustment procedures is found in Bishop et al. (1975) under the topic "Pseudo-Bayes' Estimators."

A more important issue is whether the sample is "large enough" to permit any form of categorical data analysis. Here we have few guidelines. Generally, we will require estimated expected counts of at least 1.0 in any given cell and more than 5.0 in at least 80 percent of the cells. However, discrepancies between a maximum likelihood and a weighted least squares analysis may be one symptom of a sample that is too small for either of these procedures, just as observed zeros may be.

2.8 MODEL TESTING AND BUILDING

Sections 2.1–2.7 focused on building probability models in order to characterize a process that led to an observed data set. This led to several distributions that are useful in categorical data analysis. An alternative approach makes assumptions about the underlying probability structure and then compares and contrasts various data sets. For example, the differences in death rates among various economic or social groups may indicate different levels of health status. If this can be confirmed through detailed analysis, specific health programs can be implemented. Similarly, the analysis of rates over several time periods can suggest the existence of trends. Such trends may be the basis for projections of future patterns.

We will examine two sets of infant mortality rates in this section. The first example uses goodness-of-fit tests to group U.S. census divisions into regions with similar infant mortality rates. Maximum likelihood estimates can be computed with an inexpensive pocket calculator. In the second example the infant mortality rates for New England from 1974 to 1981 are used to project the rate for 1982. The method of weighted least squares is used and the computations are performed using standard software (SAS Institute Inc. 1982b). Haberman (1978, pp. 43–60) discusses a maximum likelihood approach to the second example. The complete table of infant deaths and live births by census division and year is shown in Table 2A.1 in the Appendix to this chapter.

The presentation of tables of rates yields the maximum amount of data, but often such tables fail to convey to readers the clearest set of conclusions about the data. For this reason, effective summaries of complex tables can be more useful than the basic table of data. Table 2.11 contains the basic data on infant mortality in the United States for 1981. The total live births in thousands, total deaths under 1 year of age, and the rate per 1000 live births are shown in columns (1), (2), and (3). The data are reported according to the four regions, nine divisions, and the whole country. Three questions come to mind. First, is there any statistically significant variation across the country with either the regional or the divisional tabulation? Second, do we require the details of the nine divisions, or is the information provided by the regional estimates adequate to summarize the data? Put differently, are

Table 2.11 Infant Deaths and Death Rates by Division, United States, 1981

Region	Division[a]	Births (1000s) (1)	Deaths (2)	Rate/1000 (3)	Expected Deaths: Total (4)	Expected Deaths: Four-Region Model (5)	Expected Deaths: Final Model (6)
East		662.0	7,545	11.4	7,874.9	7,545	—
	New England	164.2	1,585	9.7	1,953.3	1,871.4	1,674.8
	Middle Atlantic	497.8	5,960	12.0	5,921.6	5,673.6	5,965.3
South		1,242.9	16,469	13.3	14,785.0	16,469	—
	South Atlantic	550.3	7,643	13.9	6,546.1	7,291.7	7,569.9
	E. South Central	236.7	3,183	13.4	2,815.7	3,136.4	3,256.1
	W. South Central	455.9	5,643	12.4	5,423.2	6,040.9	5,463.2
Central		933.6	11,013	11.8	11,105.7	11,013	—
	E. North Central	648.7	7,773	12.0	7,716.7	7,652.2	7,773.5
	W. North Central	284.9	3,240	11.4	3,389.1	3,360.8	3,414.0
West		796.8	8,217	10.3	9,478.4	8,217	—
	Mountain	231.9	2,388	10.3	2,758.6	2,391.5	2,365.3
	Pacific	564.9	5,829	10.3	6,719.8	5,825.5	5,761.9
Total		3,635.3	43,244	11.9	43,244	43,244	43,244

[a]New England: Connecticut, Maine, Massachusetts, New Hampshire, Rhode Island, Vermont; Middle Atlantic: New Jersey, New York, Pennsylvania; South Atlantic: Delaware, District of Columbia, Florida, Georgia, Maryland, North Carolina, South Carolina, Virginia, West Virginia; E. South Central: Alabama, Kentucky, Mississippi, Tennessee; W. South Central: Arkansas, Louisiana, Oklahoma, Texas; E. North Central: Illinois, Indiana, Michigan, Ohio, Wisconsin; W. North Central: Iowa, Kansas, Minnesota, Missouri, Nebraska, North Dakota, South Dakota; Mountain: Arizona, Colorado, Idaho, Montana, Nevada, New Mexico, Utah, Wyoming; Pacific: Alaska, California, Hawaii, Oregon, Washington.

the divisions homogeneous within the four regions? Finally, can the divisions be rearranged into more homogeneous regions?

Statistical answers to these questions require a few assumptions on which to base appropriate tests. We begin by formulating the outcome variable as being the death of an infant. The first assumption is that infant deaths reported in a given year occur to infants born in the same year. Since the infants dying later in life are more likely to die in a year different from the year of their birth, this is somewhat tenuous. This problem is known as one of "separation factors" and is discussed in great detail in many demography texts, such as that by Shyrock and Siegel (1973). However, if the infant mortality and birth rates are reasonably stable, the assumption is approximately true.

Next, we assume that the chances of death within any one division are relatively homogeneous. In the absence of more detailed information, this is a necessary assumption. Similarly, we assume that a live birth ending in death before 1 year of age is independent of any other such birth outcome. Finally, we assume that the chance of an infant death is small and that there are many births to be considered. Since births are numbered in the hundreds of thousands and less than 2 percent of these lead to an infant death, this is a reasonable assumption. Homogeneity and independence are more tenuous but not unreasonable. However, as a "first approximation," infant mortality may be viewed as a Poisson process, a model discussed in greater detail by Chiang (1968). Using this model, we can answer the three questions.

The first question asks about the variation of the regions and divisions from the national rate of 11.9 per 1000 live births. This corresponds to the hypotheses that the divisional and regional rates are in fact equal to the national rate. Let r_N, r_{Ri}, and r_{Dj} denote the underlying national, regional, and divisional Poisson parameters, where $i = 1, 2, 3, 4$ and $j = 1, 2, \ldots, 9$. The hypotheses are then given by:

Regional homogeneity: $H_{\text{OR}}: r_{R1} = r_{R2} = r_{R3} = r_{R4} = r_N;$

Divisional homogeneity: $H_{\text{OD}}: r_{D1} = r_{D2} = \cdots = r_{D9} = r_N.$

We can test these hypotheses by applying any of our goodness-of-fit statistics from Section 2.6. First we estimate r_N. The maximum likelihood estimate is given by $\hat{r}_N$ = deaths/births = 43,244/3635.3 = 11.9. We apply this estimate to the total live births by region and division to estimate expected infant deaths. The results are shown in column (4) of Table 2.11. Thus we have the observed counts, y_i, and expected counts given $\hat{r}_N$. Using the Pearson chi-square, we can test the regional homogeneity hypothesis:

$$X^2 = \frac{\Sigma_{i=1}^{I} (y_i - \hat{m}_i)^2}{\hat{m}_i}$$

$$= \frac{(7545 - 7874.9)^2}{7874.9} + \frac{(16{,}469 - 14{,}785.0)^2}{14{,}785.0}$$

$$+ \frac{(11{,}013 - 11{,}105.7)^2}{11{,}105.7} + \frac{(8217 - 9478.4)^2}{9478.4}$$

$$= 13.82 + 191.81 + 0.77 + 167.87 = 374.27.$$

The degrees of freedom are the number of categories (4) less the number of estimated parameters: df = 4 − 1 = 3. Notice that this is the same as the number of distinct hypotheses we are testing: $r_{R1} = r_{R2}$, $r_{R1} = r_{R3}$, $r_{R1} = r_{R4}$. Clearly, the regions are not homogeneous. Considering the contribution by each region to the X^2 statistic, it is evident that the West is very much below the national infant mortality rate and the South exceeds it. A similar computation yields the Neyman chi-square (Q = 381.04) and the likelihood ratio chi-square (G^2 = 375.56).

Since the regions are composed of divisions, it should be evident that they cannot be homogeneous. The computation of the three chi-square statistics confirms this: X^2 = 485.17, Q = 495.12, G^2 = 486.52. The degrees of freedom are computed based on having nine categories: df = 9 − 1 = 8, which corresponds to the number of distinct hypotheses.

The second question is whether the divisions are homogeneous within each of the four regions, or the hypothesis of intraregion homogeneity:

$$\begin{aligned} H_0:\ & r_{D1} = r_{D2} = r_{R1} && \text{(East)} \\ & r_{D3} = r_{D4} = r_{D5} = r_{R2} && \text{(South)} \\ & r_{D6} = r_{D7} = r_{R3} && \text{(Central)} \\ & r_{D8} = r_{D9} = r_{R4} && \text{(West)} \end{aligned}$$

To obtain expected deaths we estimate the r_{Ri} in the same manner as before: East: $r_{R1} = 7545/662.0 = 11.4$; South: $r_{R2} = 16{,}469/1242.9 = 13.3$; Central: $r_{R3} = 11{,}013/933.6 = 11.8$; West: $r_{R4} = 8217/796.8 = 10.3$. These estimated rates are then applied to the live birth totals in each division to obtain the expected deaths given the hypothesis of intraregional homogeneity. The results are shown in column (5) of Table 2.11. Notice that the expected regional totals equal the observed totals.

We can now use the goodness-of-fit statistics to test the intraregion hypothesis. The results are shown in Table 2.12. The degrees of freedom reflect the estimation of four parameters: df = 9 − 4 = 5, or the testing of five distinct hypotheses. Again, the hypothesis is rejected, suggesting significant heterogeneity within the regions.

Before moving to the third question, notice that in Table 2.12 the likelihood ratio chi-square statistics for the hypotheses of regional and intraregional homogeneity add up to the statistic for divisional homogeneity, except for rounding error. It can be shown that when maximum likelihood estimators are used, this type of partitioning will always occur. We will observe a similar result for the Neyman chi-square when weighted least squares estimators are used. Note also that while the Pearson and Neyman chi-squares with maximum likelihood estimators do not quite achieve the simple arithmetic of the likelihood ratio partition, the agreement is close. Finally, note that all the goodness-of-fit statistics are similar, but the Neyman is systematically the largest.

We can now attempt to formulate a more homogeneous grouping of

Table 2.12 Tests of Regional and Divisional Homogeneity Hypotheses

		Chi-Squares		
Hypothesis	df	Pearson	Neyman	Likelihood Ratio
Divisional homogeneity	8	485.17	495.12	486.54
Regional homogeneity	3	374.27	381.04	375.56
Intraregional homogeneity	5	108.37	116.79	110.93
Proposed model	6	22.95	23.53	23.14

the divisions than provided by the four-region model (question 3). An approach to this is to adapt the method of least significant differences used in analysis of variance (e.g., Kleinbaum and Kupper 1978, pp. 265–268). The idea is to compute all pairwise z-tests, then declare as significant those that exceed an appropriate percentile of the standard normal distribution. The percentile is determined so as to adjust for the number of tests. In the present example there are nine regions, so there are ${}_9C_2 = 36$ pairwise comparisons and significance can be declared for positive or negative results (two-tailed tests). Hence to preserve an overall level we use the $(1 - \alpha/72)100$ percentile. For example, set $\alpha = 0.05$; the percentile is $(1 - 0.05/72)100 = 99.931$, which corresponds to $z = 3.20$. So to preserve an overall $\alpha = 0.05$ significance test, a z-test must exceed 3.20 in absolute value.

The pairwise z-tests are constructed on the natural logarithmic scale so as to improve the normal approximation (see Section 2.5). Let $\hat{L}_{ij} = \ln(\hat{r}_i) - \ln(\hat{r}_j)$; then the standard error of $\hat{L}_{ij}$ is given by

$$\mathrm{SE}(\hat{L}_{ij}) = \mathrm{Var}[\ln(\hat{r}_i)] + \mathrm{Var}[\ln(\hat{r}_j)] = \left(\frac{1}{D_i} + \frac{1}{D_j}\right)^{1/2}$$

where D_i is the number of deaths in the ith division. The Z-test is $Z_{ij} = L_{ij}/\mathrm{SE}(L_{ij})$. For example, to compare New England ($\hat{r}_1 = 9.7$) to Mountain ($\hat{r}_8 = 10.3$), we compute

$$\hat{L}_{18} = \ln\frac{1585}{164.2} - \ln\frac{2388}{231.9} = -0.065,$$

$$\mathrm{SE}(\hat{L}_{18}) = \left(\frac{1}{1585} + \frac{1}{2388}\right)^{1/2} = 0.032 \qquad z_{18} = \frac{-0.065}{0.032} = -2.00.$$

So given the 72 possible comparisons, the New England and Mountain divisions do not have significantly different infant mortality rates.

The complete set of comparisons is shown in Table 2.13. Here the divisions have been rearranged in ascending order of the rates. The upper triangle shows the values of $\hat{L}_{ij}$ and $\mathrm{SE}(\hat{L}_{ij})$, and the lower triangle shows the corresponding z-tests. For example, comparing the Mountain division to the Middle Atlantic division yields $L_{ij} = 0.151$, $\mathrm{SE}(L_{ij}) = 0.024$, $z = 6.22$. The z-statistic is greater than 3.20, so the two divisions are significantly different at the overall $\alpha = 0.05$ level.

Table 2.13 Natural Logarithms of Infant Mortality Rates and Standard Errors by Division of United States, 1981[a]

Region:[b]	NE	MT	PA	WNC	MA	ENC	WSC	ESC	SA
Rate:	9.7	10.3	10.3	11.4	12.0	12.0	12.4	13.4	13.9
NE	L_{ij}[c]	0.065	0.067	0.164	0.215	0.216	0.249	0.332	0.364
	SE(L_{ij})[d]	0.032	0.028	0.031	0.038	0.028	0.028	0.031	0.028
MT	2.00	—	0.002	0.099	0.151	0.152	0.184	0.267	0.299
			0.024	0.027	0.024	0.023	0.024	0.027	0.023
PA	2.35	0.08	—	0.097	0.149	0.149	0.182	0.265	0.297
				0.022	0.018	0.017	0.019	0.022	0.017
WNC	5.35	3.68	4.44	—	0.051	0.052	0.085	0.168	0.200
					0.022	0.021	0.022	0.025	0.021

MA	5.70	6.22	8.07	2.36	—	0.001	0.033	0.116	0.148
						0.017	0.019	0.122	0.017
ENC	7.84	6.48	8.63	2.50	0.05	—	0.032	0.115	0.148
							0.017	0.021	0.016
WSC	8.75	7.54	9.74	3.84	1.79	1.86	—	0.083	0.115
								0.022	0.018
ESC	10.78	9.86	12.02	6.72	5.29	5.48	3.74	—	0.032
									0.021
SA	13.18	12.76	17.09	9.54	8.59	9.17	6.56	1.53	—

[a]Upper triangle gives log rates and standard errors; lower triangle gives pairwise z-test.
[b]NE, New England; MT, Mountain; PA, Pacific; WNC, West North Central; MA, Middle Atlantic; ENC, East North Central; ESC, East South Central; SA, South Atlantic.
[c]$\hat{L}_{ij} = \ln(r_j) - \ln(r_j)$.
[d]$SE(\hat{L}_{ij}) = \left(\frac{1}{D_i} + \frac{1}{D_j}\right)^{1/2}$, $Z = \hat{L}_{ij}/SE(\hat{L}_{ij})$.

A line is drawn on Table 2.13 separating the significant from nonsignificant comparisons. This results in four groups:

Group I: New England, Mountain, Pacific
Group II: West North Central, Middle Atlantic, East North Central
Group III: Middle Atlantic, East North Central, West South Central
Group IV: East South Central, South Atlantic.

Groups I and IV represent the lowest and highest infant mortality rates and do not overlap any other groups. The remaining two groups are more problematic since the Middle Atlantic and East North Central divisions appear in both. There are a number of possibilities, including simply lumping groups II and III together. If we do this, the corresponding goodness-of-fit statistics are shown in column (6) of Table 2.11. Although this indicates a statistically significant lack of fit ($X^2 = 22.95$, df $= 6$, $p = 0.001$), it is much better than the original four-region model. Moreover, if the West North Central division is separated out, a nearly perfect fit is obtained.

In conclusion, there is significant heterogeneity in infant mortality both within and among the usual four regions. However, an alternative grouping of the divisions into three or four regions does seem to provide an adequate summary of the data. Finally, there seems to be three basic infant mortality rates prevailing: low: 10, medium: 12, and high: 13+ per 1000 live births. The comparison of infant mortality rates across regions leads to an analysis that is guided by a series of hypotheses about regional homogeneity. These hypotheses are entirely analogous to the usual hypotheses in analysis of variance, and the fitted models can be described in terms of analysis-of-variance models. This analogy will be developed in detail when more complex categorical data problems are considered. But before that we conclude by analyzing data for trends in order to make a projection about future infant mortality rates in New England.

The data in Table 2.14 show the live births, infant deaths, and infant mortality rates for New England from 1974 to 1981. A plot of the rates or the natural logarithms of the rates suggests a nearly linear decline over the period. The usual linear model for this would be $L_i = \ln(r_i) = B_0 + B_1X_i + e_i$, where r_i is the infant mortality rate in year i, B_0 and B_1 are unknown regression coefficients, X_i is a variable for year i, and e_i is an error term. In this case we set $X_i = \text{year}(i) - 1978$ to simplify the computations. We could proceed with the usual regression

Table 2.14 Live Births, Infant Deaths, and Infant Mortality Rates per 1000 Live Births, New England, 1974–1981

Year	Live Births (1000s)	Infant Deaths	Infant Mortality Rate/1000	Log Rates
1974	152.7	2216	14.5	2.67
1975	146.4	2014	13.8	2.62
1976	146.2	1853	12.7	2.54
1977	152.3	1818	11.9	2.48
1978	153.7	1751	11.4	2.43
1979	158.3	1706	10.8	2.38
1980	157.3	1614	10.3	2.33
1981	164.2	1585	9.7	2.27

Source: National Center for Health Statistics (1983).

analysis except that we know the variance of e_i depends on the number of infant deaths in year i: $V[\ln(\hat{r}_i)] = 1/D_i$. An alternative is to use weighted least squares, which will yield statistics equivalent to the likelihood-based statistics for large samples (Neyman 1949, Bhapkar 1966, Grizzle et al. 1969).

The estimators of B_0 and B_1 are based on weighted averages of the L_i and X_i, where the weights are the inverses of $V(L_i)$:

$$\bar{X}_i = \frac{\Sigma_{i=1}^I D_i X_i}{\Sigma_{i=1}^I D_i} \qquad \bar{L}_i = \frac{\Sigma_{i=1}^I D_i L_i}{\Sigma_{i=1}^I D_i}. \tag{2.49}$$

The estimators are

$$b_1 = \frac{\Sigma_{i=1}^I D_i(X_i - \bar{X})(L_i - \bar{L})}{\Sigma_{i=1}^I D_i(X_i - \bar{X})^2} \qquad b_0 = \bar{L} - b\bar{X}_1. \tag{2.50}$$

Using the data from Table 2.13, we obtain

$$\bar{L} = \frac{2216(2.67) + 2014(2.62) + \cdots + 1585(2.27)}{14{,}557} = 2.48$$

$$\bar{X} = \frac{2216(-4) + 2014(-3) + \cdots + 1585(3)}{14{,}557} = -0.74.$$

The regression coefficients are estimated to be $b_1 = 2.44$ and $b_0 = -0.06$. These can be obtained from any standard weighted least squares program. In the present case the program PROC REG (SAS 1982b) was used.

The fit of the data to the model is based on the Neyman chi-square, which is the usual error sum of squares from the weighted regression analysis. In the present case there is no evidence of a significant lack of fit ($Q = 1.21$, df $= 6$). Using the regression equation, the predicted logarithmic infant mortality rate for 1982 is $L_9 = 2.44 - (0.06)4 = 2.20$, a rate of 9.1 per 1000 live births. The standard error of the estimated logarithm of the rate is 0.019, so the corresponding asymmetric 95 percent confidence interval is $\exp[2.20 \pm 1.96(0.019)] = (8.7, 9.40)$. An annotated printout of these results is shown in Figure 2A.1. Computationally, the sums of squares correspond to the Neyman chi-squares and the correct standard errors are obtained by dividing the regression standard errors by the square root of the mean square error. Finally, the computational formula for the Q statistic is

$$Q = \sum_{i=1}^{I} (\hat{L}_i - \bar{L})^2 D_i, \qquad \text{where } \hat{L}_i = b_0 + b_1 X_i \tag{2.51}$$

with df = (number of populations) − (number of estimates) = 8 − 1 = 6.

APPENDIX 2A: DETAILED DATA FOR INFANT MORTALITY ANALYSIS AND SAMPLE PROGRAM

Table 2A.1 Births and Infant Deaths by Census Division and Year, 1974–1982[a]

Census Division	Year								
	1974	1975[b]	1976	1977	1978	1979	1980	1981	1982[b]
New England	152.7	146.4	146.2	152.3	153.7	158.3	157.3	164.2	—
	2216	2014	1853	1818	1751	1706	1614	1585	1685
Middle Atlantic	485.5	475.2	474.1	487.4	479.0	481.5	484.6	497.8	505.4
	8019	7611	7406	6952	6539	6849	6461	5960	5692
E. North Central	612.3	609.8	598.8	627.2	622.6	647.7	661.3	648.7	640.2
	10,380	9736	8965	8899	8533	8326	8188	7773	7203
W. North Central	245.3	248.8	250.8	266.3	267.1	279.1	289.0	284.9	284.5
	3942	3684	3738	3572	3613	3324	3316	3240	2968
South Atlantic	496.5	482.5	481.1	509.4	506.6	528.2	547.5	550.3	563.5
	9027	8542	8130	8036	7888	7847	7730	7643	7312
E. South Central	224.9	224.3	222.6	236.3	231.9	239.0	244.6	236.7	235.4
	4343	4035	3966	3783	3548	3495	3483	3183	3175
W. South Central	362.1	365.7	373.0	389.2	395.7	426.7	435.5	455.9	439.9
	6339	6079	6016	5894	5897	5471	5434	5643	5582
Mountain	173.7	175.4	184.9	196.0	201.6	215.7	225.9	231.9	231.2
	2701	2584	2597	2415	2455	2617	2529	2388	2317
Pacific	417.7	421.9	444.9	468.2	480.1	487.9	541.2	564.9	573.4
	5915	5856	5677	5691	5803	5456	6247	5829	5695

[a]Births in thousands.
[b]Provisional statistics.
Source: National Center for Health Statistics (1983).

```
DATA MORT;
  INPUT YEAR POP DEATH;
  LNRATE = LOG(DEATH/POP);
  CODE = YEAR - 1978;
  LABEL
    LNRATE = NATURAL LOG OF INFANT MORTALITY RATE
    CODE = CENTERED YEAR
    POP = LIVE BIRTHS IN 1000S
    DEATH = INFANT DEATHS
;CARDS;
1974 152.7 2216
1975 146.4 2014
1976 146.2 1853
1977 152.3 1818
1978 153.7 1751
1979 158.3 1705
1980 157.3 1614
1981 164.2 1585
1982 .     1685
;TITLE 'NEW ENGLAND IMR ANALYSIS 1974 TO 1981';
PROC REG;
  ID YEAR;
  MODEL LNRATE = CODE/R CLI SS2;
  WEIGHT DEATH;
TITLE2 'WEIGHT (GSK) ANALYSIS';
```

Figure 2A.1 SAS code for weighted least squares analysis of ln (infant mortality rates).

3

Contingency Tables

Chapter 2 presented the basic concepts of probability distributions. This tells us something about the foundation of categorical data analysis. The rationale for understanding discrete variables is that in applications we will be comparing the probability distributions across several populations. Alternatively, when variables are measured simultaneously, questions of interest concern associations among them. These two viewpoints are referred to as hypotheses of homogeneity and independence, respectively. For investigations involving only two variables, the test statistics are generally identical but interpretations of results will differ. In Chapters 4–8, involving more than two variables the structure of the problem will lead to different analytical strategies

The study of contingency tables begins with the simple problem of two dichotomous variables or the 2×2 *table*. The problem is extended by first examining $I \times 2$ tables and then $I \times J$ tables. This terminology means that we consider tables based on two variables, where one variable has $I > 2$ levels and the other has 2 and then J levels.

After testing basic hypotheses we will use indices of association to summarize our findings. For the simplest case, the 2×2 table, this is fairly straightforward, but for more complex tables the choice depends heavily on the structure of the data set and sampling.

3.1 HYPOTHESES OF HOMOGENEITY AND INDEPENDENCE

Epidemiologists and health researchers have been interested in the relationship between job stress and coronary heart disease (CHD) for a number of years. In the Western Electric Longitudinal Study, workers were asked a variety of questions about their work environment and then followed up for a period of 10 years. This follow-up permitted the identification of major CHD outcomes. Ostfeld (1980) reported on the relationship between CHD and working under tension. The key table to his discussion is reproduced in Table 3.1; the columns correspond to whether subjects reported working under tension at baseline, and the rows indicate whether a major CHD event occurred in the ensuing 10 years. The table shows that 404 workers reported working under tension and of these 97 (24 percent) suffered a major CHD event. Compare this to the 1609 who did not report working under tension. Among these

Table 3.1 Major CHD Event by Report of Job Tension: Observed Counts and Proportions for 2013 Western Electric Workers

Do You Work Under Tension?	Major CHD Event: Yes	Major CHD Event: No	Total
Yes	97	307	404
	0.24	0.76	1.00
No	200	1409	1609
	0.12	0.88	1.00
Total	297	1716	2013
	0.15	0.85	1.00

subjects 200 or 12.4 percent suffered a similar outcome. The difference in the proportion suffering a major CHD event suggests that these are samples from distinct populations. That is, the tension question identifies heterogeneous strata in the underlying population. Moreover, those who report working under tension are almost twice as likely to have a major CHD event.

Studies where one variable is fixed and other is random are called *one-factor, one-response* problems. In the example, the presence or absence of tension measured at baseline defines the factor. We are interested in comparing the response, defined as the presence or absence of CHD after 10 years, across the levels of the factor. The null hypothesis is that the populations defined by the levels of the factor are homogeneous.

Let p_j denote the probability of a major CHD event, $j = 1, 2$, for the jth answer at baseline. The form of the null hypothesis is: H_0: $p_1 = p_2$. Two equivalent forms are H_0': $p_1 - p_2 = 0$ and H_0'': $p_1/p_2 = 1$. Clearly, if any form of the hypothesis is true, the others must be true. The second form will lead us to an appropriate test statistic in Section 3.2. The third form of the null hypothesis is often used by epidemiologists since it corresponds to the "relative risk" of the event under study. One final point about the hypothesis of homogeneity is that the sample size at each level of the factor is considered fixed, not random Thus we treat the 404 yeses and 1609 noes as fixed by design or at least prior to the CHD event.

The Western Electric Longitudinal Study is a *prospective* or *cohort study*. These studies typically begin with samples from defined populations and then follow them over time. Such studies can be expensive and are inherently time consuming. For rare diseases, such as particular types of cancer, cohort studies may not be feasible. An alternative is the *case-control* or *retrospective study*, which begins by selecting subjects with the disease and a disease-free comparison group. An example of such a study is reported by Graham and Shotz (Table 3.2). They reported that among 49 cases of cervical cancer, 7 or 14.3 percent had their first pregnancy after age 25. Among a sample of 310 controls, 107 or 34.5 percent were first pregnant after age 25. Since the sampling was based on the disease status, that corresponds to the factor. Age at first pregnancy is a dichotomous response.

The hypothesis of homogeneity corresponds to hypothesizing that

Table 3.2 Age at First Pregnancy by Cervical Cancer Among Women Aged 50–59 at Diagnosis

Age at First Pregnancy	Disease Status: Cervical Cancer	Disease Status: Controls	Total
≤25	42	203	245
	0.857	0.655	
>25	7	107	114
	0.143	0.345	
Total	49	310	359
	1.000	1.000	1.000

Source: Graham and Shotz (1979).

the probability of late first pregnancy is the same in the two groups. Let p_j denote that probability; $j = 1, 2$ for cancer and control groups. The null hypothesis is the same as before, H_0: $p_1 = p_2$. Predicting age at first pregnancy from subsequent cervical cancer is of little interest, but since the disease is rare, the odds ratio can be used to approximate the underlying relative risk. This was developed as an application of Bayes' theorem in Chapter 2. So we can rewrite our hypothesis as

$$H_0\colon \quad \frac{p_1}{1 - p_1} = \frac{p_2}{1 - p_2} \qquad \text{or} \qquad H_0\colon \quad \frac{p_1(1 - p_2)}{p_2(1 - p_1)} = 1.$$

The important point is that while the case-control study design leads to an uninteresting factor-response structure, use of the odds-ratio form of the homogeneity hypothesis allows us to examine the pertinent issues.

A third type of design is the *cross-sectional study*. It is probably the most common type of study, particularly in social science research, which relies heavily on sample surveys of large populations. In this design, samples of individuals are selected at random and then questioned about a variety of characteristics. The investigator now has less prior structure on the data and is interested in associations among the responses. An example of this is reported by Senie et al. (1981) in a study of correlates of breast self-examination (BSE) among 1216

Table 3.3 Frequency of Breast Self-Examination by Age at Diagnosis[a]

Age	Frequency of Breast Self-Examination: Monthly	Occasionally	Never	Total
<45	91	90	51	232
45–59	150	200	155	505
60+	109	198	172	479
Total	350	488	378	1216

Source: Senie et al. (1981).

women with breast cancer (Table 3.3). In this study, 232 or 19.1 percent were under age 45, 41.5 percent were age 45–59, and 39.4 percent were at least age 60 at diagnosis. For this sample 350 or 28.8 precent performed BSE monthly, 40.1 percent occasionally, and 31.1 percent reported never.

An immediate question is whether age and BSE frequency are related. It might be argued that age forms a series of strata, so it should be considered a factor. However, since age and BSE frequency were queried simultaneously, we will treat both as random. In other words, this is a two-response problem. The corresponding null hypothesis is that the variables are independent. From the definition of independence given in Chapter 2, this means that the probabilities of the joint events in the cells of Table 3.3 can be written in terms of the marginal probabilities.

Let p_{ij} denote the probability that a woman is in the ith age group and perfroms BSE with frequency j, $i = 1, 2, 3$, $j = 1, 2, 3$. That is,

$$P(\text{age} = i,\ \text{BSE} = j) = p_{ij} > 0 \qquad \text{where} \sum_{i=1}^{3} \sum_{j=1}^{3} p_{ij} = 1. \tag{3.1}$$

From the definition of conditional probability (2.7), we obtain

$$\begin{aligned} p_{ij} &= P(\text{age} = i,\ \text{BSE} = j) = P(\text{age} = i)P(\text{BSE} = j|\text{age} = i) \\ &= p_{i+}P(\text{BSE} = j|\text{age} = i). \end{aligned} \tag{3.2}$$

Here $p_{i+} = \Sigma_{j=1} p_{ij}$; the "+" subscript indicates which variable has been summed over or "collapsed." Now if BSE and age are independent, then from (2.10),

$$P(\text{BSE} = j|\text{age} = \text{i}) = P(\text{BSE} = j) = p_{+j}. \tag{3.3}$$

So the hypothesis of independence is given by

$$H_0\colon \; P(\text{age} = i, \text{ BSE} = j) = P(\text{age} = i)P(\text{BSE} = j) \tag{3.4}$$

$$\text{or} \quad H_0\colon \; p_{ij} = p_{i+}p_{+j} \qquad i = 1, 2, 3, j = 1, 2, 3.$$

For example, if age and BSE are independent, the proportion of women < 45 never practicing BSE should be $p_{1+}p_{+3} = 0.191(0.311) = 0.059$ instead of the observed value of 0.042. The relationship (3.4) will have to hold in all nine cells for the null hypothesis to be accepted. Appropriate test statistics for this will be developed in Sections 3.2–3.4.

To summarize, in many studies data are collected as responses at each of several levels of a factor. In this circumstance we are interested in testing the homogeneity of the responses across the levels of the factor. Occasionally, the factor-response structure is of no particular interest, but by making appropriate assumptions, the hypothesis can be rewritten in a more interesting form. An example of this is with case-control studies, where the odds ratio is used. In other studies all the variables are measured simultaneously. Although it is often possible to make some assumptions and convert several of the variables to factors, it may be of interest to focus on the hypothesis of independence. Fortunately, for contingency tables with only two variables the statistical tests are unaffected by the two approaches. Moreover, it can be shown that if homogeneity is true, independence is also, and conversely.

3.2 2 × 2 TABLES

3.2.1 The Prospective Study: Homogeneity

The data presented by Ostfeld (1980) and shown in Table 3.1 indicate that people who report working under tension have almost twice the rate of a major CHD event as those who report not working under tension (24.0 versus 12.4 percent). Aside from questions of study design and unreported variables, we have the question of whether these percentages could have arisen by chance alone. To answer this statistical question, we will need to build a probability model and then assess whether the observed outcome is reasonable under the model.

The model is in two parts. First, a series of assumptions imply there are binomial distributions within each level of the factor. Second, we assume that the hypothesis of homogeneity is true and determine how likely the outcome would be. The basic assumption is that we have drawn simple random samples of sizes 404 and 1609 from populations defined by the two levels of the factor. The simple random sampling assumption can be broken down into the "independent and identical distribution" assumptions (i.i.d.). Independence means that all 2013 subjects, included in the sample, were interviewed and followed in such a way that no subject's classification affected any other subject's classification. For example, we assume at baseline that each subject provided answers to the tension question without reference to the answer of any other subject. Similarly, we assume that subjects were all followed for 10 years with equal vigor and that any nonresponse or loss-to-follow-up occurred at random.

The second part of the i.i.d. assumption (identical distribution) is that for all workers in a given level of tension, the probability of a major CHD event is the same. That is, for the 404 workers who reported working under tension, the unknown probability of a CHD event is p_1. Similarly, for all the other workers sampled, the unknown probability is p_2. These assumptions lead to a binomial distribution for each of the two groups where the random variable, Y_j ($j = 1, 2$), is the number of events in each group.

The hypothesis of homogeneity compares the two binomial distributions implied by the assumptions. More precisely, it compares the two parameters after adjusting for different sample sizes. The form of the hypothesis is

$$H_0: \quad p_1 = p_2 = p. \tag{3.5}$$

Thus we estimate p and then determine the expected number of subjects in each cell under the hypothesis. The chi-square goodness-of-fit statistics in Section 2.6 can be used as our test statistics. Since there are only two random variables (Y_1, Y_2) and one parameter to be estimated (p), the degrees of freedom are df $= 2 - 1 = 1$.

The maximum likelihood estimate of p is $\hat{p} = (Y_1 + Y_2)/n = 297/2013 = 0.148$. The expected counts are $\hat{m}_1 = n_1\hat{p} = 404(0.148) = 59.607$, $\hat{m}_1 = n_1\hat{p} = 1609(0.148) = 237.393$. Note that $\hat{m}_1 + \hat{m}_2 = 297.000$, $n_1 - \hat{m}_1 = 344.393$, and $n_2 - \hat{m}_2 = 1371.607$. The chi-square statistics are therefore (2.46), (2.47), and (2.48):

$$X^2 = \frac{(97 - 59.607)^2}{59.607} + \frac{(200 - 237.393)^2}{273.393}$$

$$+ \frac{(307 - 344.393)^2}{344.393} + \frac{(1409 - 1371.607)^2}{1371.607}$$

$$= \frac{(37.393)^2}{59.607} + \frac{(-37.393)^2}{237.393}$$

$$+ \frac{(-37.393)^2}{344.393} + \frac{(37.393)^2}{1371.607}$$

$$= 1398.27\left(\frac{1}{59.607} + \frac{1}{237.393} + \frac{1}{344.393} + \frac{1}{1371.607}\right) = 34.43.$$

Similarly, $Q = (1398.27)(1/97 + 1/200 + 1/307 + 1/1409) = 26.95$. Finally,

$$G^2 = 2\left[97 \ln\left(\frac{97}{59.607}\right) + 200 \ln\left(\frac{200}{237.393}\right)\right.$$

$$\left. + 307 \ln\left(\frac{307}{344.393}\right) + 1409 \ln\left(\frac{1409}{1371.607}\right)\right] = 31.13.$$

The statistics are all quite large when compared to a chi-square distribution with one degree of freedom; hence we reject the hypothesis of homogeneity and conclude that the samples were selected from different populations. It follows that the persons who report working under tension are *significantly* more likely to suffer a CHD event.

The X^2 and Q statistics were written out in some detail to illustrate an interesting result for 2×2 tables. Notice that the numerators for each are constant regardless of which cell is considered. That is, y_1 −

$\hat{m}_1 = x$, $y_2 - \hat{m}_2 = -x$, $n_1 - y_1 - (n_1 - \hat{m}_1) = -(y_1 - \hat{m}_1) = -x$, $n_2 - y_2 - (n_2 - \hat{m}_2) = -(y_2 - \hat{m}_2) = x$. Hence these chi-square statistics can be written

$$X^2 = x^2\left(\frac{1}{\hat{m}_1} + \frac{1}{\hat{m}_2} + \frac{1}{n_1 - \hat{m}_1} + \frac{1}{n_2 - \hat{m}_2}\right) \tag{3.6a}$$

$$Q = x^2\left(\frac{1}{y_1} + \frac{1}{y_2} + \frac{1}{n_1 - y_1} + \frac{1}{n_2 - y_2}\right) \tag{3.6b}$$

From this it should be clear that degrees of freedom can also be determined by noting that there is only one distinct deviation, $x = y_1 - m_1$.

Having determined that the disease rates are different, we can go further and put a confidence interval on the difference. Recall from Chapter 2 that the unbiased estimators of p_j, $j = 1, 2$ are $\hat{p}_j = Y_j/n_j$ with variance estimators $V(\hat{p}_j) = \hat{p}_j(1 - \hat{p}_j)/n_j$. This gives the $(1 - \alpha)100$ percent confidence interval for the difference as $(\alpha = 0.05)$.

$$(\hat{p}_1 - \hat{p}_2) \pm z_{\alpha/2}\sqrt{\frac{\hat{p}_1(1 - \hat{p}_1)}{n_1} + \frac{\hat{p}_2(1 - \hat{p}_2)}{n_2}}$$

$$= \frac{97}{404} + \frac{200}{1609} \pm 1.96\sqrt{\frac{97(307)}{307^3} + \frac{200(1409)}{1609^3}}$$

$$= 0.116 \pm 1.96\sqrt{0.000519}.$$

or (0.071, 0.160). That is, we are 95 percent confident that the difference is between 7.1 and 16.0 percent.

An alternative representation of the comparison is the relative risk: $RR = p_1/p_2$. From Chapter 2 recall that the variance estimator of the $\ln(\hat{p}_i)$ is given by

$$V[\ln(\hat{p}_i)] = \frac{V(\hat{p}_i)}{(\hat{p}_i)^2} = \frac{\hat{p}_i(1 - \hat{p}_i)/n}{(\hat{p}_i)^2} = \frac{1 - \hat{p}_i}{n_i\hat{p}_i}$$

The variance estimator of ln (RR) is therefore

$$V\left[\ln\left(\frac{\hat{p}_1}{\hat{p}_2}\right)\right] = \frac{n_1 - y_1}{n_1 y_1} + \frac{n_2 - y_2}{n_2 y_2}$$

Using this estimator, the 95 percent confidence interval is

$$\exp\left\{\ln\left[\frac{97}{404}\left(\frac{1609}{200}\right)\right] \pm 1.96\sqrt{\frac{307}{404(97)} + \frac{1409}{1609(200)}}\right\}$$

$$= \exp[\ln(1.93) \pm 1.96\sqrt{0.0122}]$$

or we are 95 percent confident that the relative risk of working under tension is in the interval (1.56, 2.40). That is, those reporting yes to working under tension had an increase of risk between 56 and 140 percent.

3.2.2 The Cross-Sectional Study: Independence

The analysis in Section 3.2.1 was based on comparing two samples. Suppose for the moment that the study had been conducted cross-sectionally, perhaps with different variables but the same outcomes. Then the design would have had a two-response structure and the hypothesis would have been written in terms of independence:

$$H_0: \; p_{ij} = p_{i+}p_{+j} \qquad i = 1, 2, j = 1, 2. \tag{3.7}$$

where i indexes the occurrence of a CHD event and j indexes working under tension, both taking values 1 or 2 corresponding to yes or no. Now we are in effect assuming that the data arose from a multinomial with four levels. The random variables and underlying parameters are shown in Table 3.4. Here Y_{ij} are the frequencies in each cell and p_{ij} are the multinomial parameters together with the fixed sample size. Notice there are three random variables since $\Sigma_{i=1}^{2}\Sigma_{j=1}^{2} Y_{ij} = n$ is a constraint given the fixed sample size.

The next step is to estimate the expected counts under the null hypothesis (3.7). The maximum likelihood estimators of the marginal parameters are given by $\hat{p}_1 = Y_{1+}/n$, $\hat{p}_{2+} = 1 - \hat{p}_1$, $\hat{p}_{+1} = Y_{+1}/n$, and $\hat{p}_{+2} = 1 - \hat{p}_{+1}$. There being two estimators and one constraint, this leaves $4 - 2 - 1 = 1$ degree of freedom for the test statistic. Under the null hypothesis the cell parameters are estimated from $\hat{p}_{ij} = \hat{p}_{i+}\hat{p}_{+j}$ and the expected frequencies are estimated by $\hat{m}_{ij} = n\hat{p}_{ij} = n\hat{p}_{i+}\hat{p}_{+j} = Y_{i+}Y_{+j}/n$. The results for the CHD-tension example are

Table 3.4 Multinomial Representation of a 2 × 2 Table

Responses		B		
	Level	1	2	Total
A	1	Y_{11}	Y_{12}	Y_{1+}
		p_{11}	p_{12}	p_{1+}
	2	Y_{21}	Y_{22}	Y_{2+}
		p_{21}	p_{22}	p_{2+}
	Total	Y_{+1}	Y_{+2}	n
		p_{+1}	p_{+1}	1.0

$$\hat{m}_{11} = \frac{297(404)}{2013} = 59.607 \qquad \hat{m}_{12} = \frac{297(1609)}{2013} = 237.393$$

$$\hat{m}_{21} = \frac{1716(404)}{2013} = 344.393 \qquad \hat{m}_{22} = \frac{1716(1609)}{2013} = 1371.607$$

Note that the total over all four cells is 2013. These are identical to what was obtained under the null hypothesis of homogeneity, so the test statistics must be the same. This means that we reject the hypothesis of independence and conclude that working under tension and CHD are related.

The last question is: How do we summarize this dependency? The most common method is in terms of the estimated odds ratio:

$$\hat{\text{OR}} = \frac{\hat{p}_{11}\hat{p}_{22}}{\hat{p}_{12}\hat{p}_{21}} = \frac{Y_{11}Y_{22}}{Y_{12}Y_{21}}. \tag{3.8}$$

In the present example the result is $\hat{\text{OR}} = 97(1409)/307(200) = 2.23$. The confidence interval is based on the estimated variance of $\ln(\hat{\text{OR}})$:

$$V[\ln(\hat{\text{OR}})] = \sum_{i=1}^{2} \sum_{j=1}^{2} \frac{1}{Y_{ij}} \tag{3.9}$$

For example, the 95 percent confidence interval is given by

$$\exp\left[\ln(\hat{\mathrm{OR}}) \pm 1.96\sqrt{\sum_{i=1}^{2}\sum_{j=1}^{2}\frac{1}{Y_{ij}}}\right]$$

$$= \exp\left[\ln(2.23) \pm 1.96\sqrt{\frac{1}{97}+\frac{1}{200}+\frac{1}{307}+\frac{1}{1409}}\right]$$

$$= \exp[\ln(2.23) \pm 1.96\sqrt{0.0192}].$$

Using this, we are 95 percent confident that the true odds ratio lies in the interval (1.70, 2.92). This means that we are 95 percent sure that the people reporting tension are 1.70 to 2.92 more likely to experience a major CHD event than those not reporting tension.

The results based on the two hypotheses (3.5) and (3.7) lead to the same statistical conclusion, but interpretations differ. From the prospective study we conclude that there is an increase in risk associated with the factor levels, but from the cross-sectional study we can only infer the existence of an association. Moreover, when the data are analyzed prospectively and the relative risk is used to summarize the findings, the resulting confidence interval is different from that based on the odds ratio, (1.56, 2.40) and (1.70, 2.92), respectively. Why did this discrepancy occur, and which is right? The discrepancy occurs because the rare disease assumption was violated. In the example nearly 15 percent of the subjects experienced a major CHD event, and this is too great to ensure the rare disease assumption. Hence there is a difference in the two summary measures.

The second question, Which is right?, is more difficult. As noted by many authors, both the relative risk (or simple difference) and the odds ratio have simple interpetations. The former compares the probabilities of the event, while the latter compares the odds of the events. Mathematical statistics provides no guidance since both summary measures have similar mean square errors. So which should be used? The data were collected prospectively, so it is clear that the hypothesis of homogeneity is more descriptive of the analysis. This suggests that the relative risk (or simple difference) is more appropriate. In other situations, such as cross-sectional studies, the choice is less obvious and can be left to the subjective preferences of the analyst. His or her obligation is to make clear why one or the other was utilized. Finally, in some situations, such as retrospective studies, the factor-response structure is the reverse of what is of interest. Moreover, usually the relative risk cannot be estimated directly from such studies. In this circumstance the odds ratio is the measure of choice. If the "rare disease assumption" is appropriate, the resulting estimate will be close to the relative risk; this may simplify interpretation of the analysis.

3.2.3 The Case-Control or Retrospective Study: Odds Ratios

The reversal of the factor-response structure of a case-control study may be illustrated by considering the role of early pregnancy in cervical cancer. The data shown in Table 3.2 are typical of those obtained in case-control studies of this disease. Here 49 women aged 50–59 at diagnosis of cervical cancer are compared to 310 controls. The null hypothesis is that the two groups are similar with respect to their childbearing experience. However, the null hypothesis of interest is Women have similar risks of cervical cancer regardless of their age at first pregnancy. The test statistic is the same as in the prospective study, but the summary measure is the odds ratio. The disease is fairly rare, so the estimated odds ratio should be a good estimate of the relative risk as well.

The analysis begins with a statement of the required assumptions. That is, we assume:

1. The data were independently obtained from each subject.
2. Within each group the probability of a first pregnancy after age 25 is constant.
3. The two groups are of fixed sample size.

These assumptions lead again to the two-binomial model.

Let Y_1 denote the number of cases with an early pregnancy and Y_2 the corresponding number of controls. The binomial parameters are p_1 and p_2, respectively. The null hypothesis is H_0: $p_1 = p_2 = p$. Therefore, the only parameter to be estimated under H_0 is $\hat{p} = (Y_1 + Y_2)/n = 245/359$. Next we compute the expected counts:

$$\hat{m}_1 = \frac{49(245)}{359} = 33.44 \qquad n_1 - \hat{m}_1 = 15.56$$

$$\hat{m}_2 = \frac{310(245)}{359} = 211.56 \qquad n_2 - \hat{m}_2 = 98.44.$$

The resulting chi-square test statistics are $X^2 = 7.99$, $Q = 13.26$, $G^2 = 9.04$. Since there is only one degree of freedom, we reject the null hypothesis of homogeneity and conclude that the samples have been drawn from different populations.

A simpler computing formula for the Pearson chi-square statistic is

$$X^2 = \frac{n(Y_{11}Y_{22} - Y_{12}Y_{21})^2}{Y_{1+}Y_{2+}Y_{+1}Y_{+2}}. \tag{3.10}$$

For example, $X^2 = 359[42(107) - 7(203)]^2/[49(310)(245)(114)] = 7.99$.

The strength and direction of the association is given by the odds-ratio approximation to the relative risk: $\hat{OR} = [Y_1(n_2 - Y_2)]/[Y_2(n_1 - Y_2)] = [42(107)]/[203(7)] = 3.16$. This means that women with a first pregnancy at or before age 25 are more than three times as likely to develop cervical cancer as women with later first pregnancies, at least in this population. How confident are we of this point estimate? We are 95 percent confident that the true risk is in the following interval: $\exp[\ln(3.16) \pm 1.96\sqrt{1/42 + 1/7 + 1/203 + 1/107}] = (1.37, 7.28)$. This is a fairly wide interval, so we should be somewhat cautious about our findings.

3.3 $I \times 2$ TABLES

Tables where one variable has more than two levels are a straightforward extension of our analysis of 2×2 tables. The general form of the test statistics remains the same. There are also some simple computational formulas, which are particularly interesting in illuminating the difference between the Neyman and Pearson statistics. Again we will be concerned about the underlying study design and its associated factor-response structure. Fortunately, the test statistics are generally unaffected by this structure. Finally, we consider the special hypotheses that are of interest when the I-level or polytomous variable is ordinal.

3.3.1 Nominal Variables

As an example, consider the data on the relationship between social class and major depression shown in Table 3.5. Here 22 of 511 respondents were diagnosed as suffering from major depression. The distribution of these 22 respondents is 4, 13, and 5 in the categories lower, middle, and upper class, respectively. The analysis begins by assuming a predetermined sample size of 511 respondents selected at random. That is, the responses were independently elicited. Since the survey was conducted cross-sectionally, the variables correspond to a two-response structure. The null hypothesis is that social class and diagnosis of major depression are independent.

The statistical formulation of the hypothesis begins by letting $p_{ij} > 0$ denote the unknown probability that a subject is in social class i and diagnosis category j. Here $i = 1$ for lower class, $i = 2$ for middle class and $i = 3$ for upper class; and $j = 1$ for major depression and $j = 2$ for other. The final assumption is the p_{ij} are constant for all 511 subjects and $\Sigma_{ij} p_{ij} = 1$. The null hypothesis of independence between social class and diagnosis is

$$H_0\colon \quad p_{ij} = p_{i+}p_{+j} \qquad \text{for all } i, j \tag{3.11}$$

Table 3.5 Frequency and Expected Counts of Major Depression by Social Class

Social Class	Diagnosis: Major Depression	Diagnosis: Other	Total	Proportion Major Depression
Lower	4	159	163	0.0245
	7.02	155.98		
Middle	13	212	225	0.0578
	9.69	215.31		
Upper	5	118	123	0.0407
	5.30	117.70		
Total	22	489	511	0.0431

Source: Abstracted from Weissman and Myers (1978).

where $\Sigma_i p_{ij} = p_{+j}$ and $\Sigma_j p_{ij} = p_{i+}$. To estimate the $m_{ij} = np_{ij}$ under the null hypothesis, the marginal proportions p_{i+} and p_{+j} must be estimated. Let Y_{ij} denote the observed frequencies. The maximum likelihood estimators are given by

$$\hat{p}_{i+} = \frac{Y_{i+}}{n} \qquad \text{and} \qquad \hat{p}_{+j} = \frac{Y_{+j}}{n}. \tag{3.12}$$

Since $\Sigma_i p_{i+} = 1$ and $p_{+1} + p_{+2} = 1$, only $I - 1$ and 1 of the estimators in (3.12) are independent. There is also the overall requirement that the total of the expected counts equals the total sample size. This means that the degrees of freedom are given by df $= 2I - (I - 1 + 1) - 1 = I - 1 = 3 - 1 = 2$. Alternatively, observe that (3.11) corresponds to $I - 1$ or 2 distinct null hypotheses.

The expected counts are shown as the second entries in Table 3.5. For example, $\hat{m}_{11} = 511(163/511)(22/511) = 511(0.043)(0.319) = 7.02$. The test statistics are given by

$$X^2 = \frac{(4 - 7.02)^2}{7.02} + \frac{(159 - 155.98)^2}{155.98} + \cdots + \frac{(118 - 117.70)^2}{117.70}$$

$$= 2.56$$

$$Q = \frac{(4 - 7.02)^2}{4} + \frac{(159 - 155.98)^2}{159} + \cdots + \frac{(118 - 117.70)^2}{118}$$

$$= 3.21$$

$$G^2 = 2\left[4 \ln\left(\frac{4}{7.02}\right) + 159 \ln\left(\frac{159}{155.98}\right) + \cdots + 118 \ln\left(\frac{118}{117.70}\right)\right]$$

$$= 2.49.$$

Comparing these with a chi-square distribution with two degrees of freedom, we note that the p-value is greater than 0.20, so the null hypothesis cannot be rejected. It must be concluded for this sample that social class and major depression are independent.

An alternative formulation of the design would be to consider social class as an ascribed characteristic. Then the hypothesis would be framed in terms of the homogeneity of the probabilities of major depression across social classes. That is, let p_i denote the probability of major depression in the ith social class and n_i the corresponding sample size. The homogeneity hypothesis is given by

$$H_0\colon\quad p_1 = p_2 = p_3 = p. \tag{3.13}$$

We now need to estimate p. The maximum likelihood estimate is

$$\hat{p} = \frac{Y_{+1}}{n} = \frac{22}{511} = 0.043. \tag{3.14}$$

This is applied to each of the samples to obtain the estimated expected counts: $\hat{m}_{i1} = \hat{p}n_i$ and $\hat{m}_{i2} = n_i - \hat{m}_{i1}$. For example, $\hat{m}_{21} = 0.043(225) = 9.69$. It can be shown that the expected counts under the hypothesis of homogeneity are identical to those under independence and hence the chi-square statistics are identical.

The formulation of the hypothesis in terms of homogeneity does suggest a simpler computational formula, as noted by Snedecor and Irwin (1933):

$$X^2 = \frac{\sum_{i=1}^{I} (\hat{p}_i - \hat{p})^2 n_i}{\hat{p}(1 - \hat{p})} \tag{3.15}$$

where $\hat{p}$ and $\hat{p}_1, \hat{p}_2, \ldots, \hat{p}_I$ are estimated in the usual manner. In the example, these are shown in the last column of Table 3.5. The resulting Pearson chi-square computation is

$$X^2 = \frac{163(0.0245 - 0.0431)^2 + 225(0.0578 - 0.0431)^2 + 123(0.0407 - 0.0431)^2}{0.0431(0.9569)}$$

$$= 2.56.$$

An interesting aspect of equation (3.15) is that the coefficients of the squared differences are precisely the inverse of the standard errors of the p_i under the null hypothesis. That is, the Pearson chi-square corresponds to the weighted sum of the $(p_i - p)^2$ under the null hypothesis. With a little bit of algebra it can be shown that the Neyman chi-square is given by

$$Q = \frac{\sum_{i=1}^{I} (\hat{p}_i - \hat{p})^2 n_i}{\hat{p}_i(1 - \hat{p}_i)}. \tag{3.16}$$

This means that it is a weighted sum also, but the weights are now based on the estimated variances under the alternative hypothesis.

A final point about the weighted structure of these chi-square statistics regards the estimator of p. As we noted earlier, the maximum likelihood estimator is given by (3.14), which can be written as

$$\hat{p} = \frac{Y_{11} + Y_{21} + \cdots + Y_{I1}}{n}$$

$$= \frac{(n_1 Y_{11}/n_1) + (n_2 Y_{21}/n_2) + \cdots + (n_I Y_{I1}/nI)}{n}$$

$$= \frac{n_1 p_1 + n_2 p_2 + \cdots + n_I p_I}{n}$$

Thus p is the weighted average of the p_i, where the weights are the corresponding sample sizes. The corresponding weighted least squares estimator is given by

$$\tilde{p} = \frac{\sum_{i=1}^{I} v_i^{-1} p_i}{\sum_{i=1}^{I} v_i^{-1}} \tag{3.17}$$

where $v_i = \hat{p}_i(1 - \hat{p}_i)/n_i$. It can be shown that this estimator is equivalent to the maximum likelihood estimator for large samples. For two-way tables the maximum likelihood estimate is preferred because it is unbiased and easier to compute. The latter reason is probably more important.

The data on social class and major depression yielded chi-square statistics less than 3.84, the 95th percentile of the chi-square distribution with one degree of freedom. This suggests that there is no single component of the chi-square that could be statistically significant. This not always the case, as we shall see in the next section.

3.3.2 Tests for Trend

Sometimes in an $I \times 2$ table the polytomous (I-level) variable is ordinal. An example of such a variable is the degree of improvement experienced by patients after treatment, as in the data displayed in Table 3.6. Here leprosy patients with different levels of disease, little or much skin infiltration, were graded according to their improvement after treatment. Details about the study are found in Cochran (1954) and Snedecor and Cochran (1967). Here it is evident that the degree of infiltration is a two-level factor and change in health is a five-level ordinal response. We will index group by $i = 1$ or 2 and change in health by $j = 1, 2, 3, 4, 5$.

There are two questions that come to mind. First, are the two response distributions homogeneous? Second, is the degree of im-

Table 3.6 196 Leprosy Patients Classified According to Change in Health and Degree of Skin Infiltration

		Degree of Infiltration			
Change in Health	Score	Little Y_{i1} p_{i1} (1)	Much Y_{i2} p_{i1} (2)	Total Y_{i+} p_{i+} (3)	Proportion Much Infiltration q_i (4)
Improvement					
Marked	3	11	7	18	0.3889
		0.076	0.135	0.092	
Moderate	2	27	15	42	0.3571
		0.188	0.288	0.214	
Slight	1	42	16	58	0.2759
		0.292	0.308	0.296	
Stationary	0	53	13	66	0.1970
		0.368	0.250	0.337	
Worse	−1	11	1	12	0.0833
		0.076	0.019	0.061	
Total Y_{+j}		144	52	196	0.2553
		1.000	1.000	1.000	

Source: Snedecor and Cochran (1984).

provement the same in the two groups? An alternative form of the second hypothesis asks whether there is a trend in the proportion of subjects who experienced such infiltration. This alternative form reverses the factor-response structure; therefore, the tests for statistical significance will have to be interpreted cautiously. However, Yates (1948) showed that the two forms of the trend hypothesis are essentially the same.

Let $p_{ij} > 0$ denote the probability that an individual in group j (= 1, 2 for little or much infiltration) experiences improvement at level i (=1, 2, 3, 4, 5 for marked, moderate, slight, stationary, worse). There are two implied constraints $\Sigma_{i=1}^{5} p_{ij} = 1$, $j = 1, 2$. The null hypothesis is given by

$$H_0\colon\ p_{i1} = p_{i2} = p_i \qquad i = 1, 2, 3, 4, 5. \tag{3.18}$$

Y_{ij} is the corresponding multinomial random variable for the number of individuals of group j classified at level i. The maximum likelihood estimators are $\hat{p}_i = Y_{i+}/n$.

The estimates are shown in column (3) of Table 3.6. The expected values of Y_{ij} are estimated by $\hat{m}_{ij}$, where $\hat{m}_{ij} = \hat{p}_i n_j$. Here n_j denotes the sample size in group j. For example, $\hat{m}_{21} = (0.2143)144 = 30.86$. Using the estimated expected values, the test statistics are $X^2 = 6.88$, $Q = 9.88$, $G^2 = 7.28$. There are 10 cells, 2 constraints, and 4 estimated parameters, so df $= 10 - 4 - 2 = 4$. The largest test statistic, Q, corresponds to the 95.8th percentile of a chi-square distribution with df $= 4$. The smallest, X^2, corresponds to the 85.8th percentile. Intermediate to these two, G^2 is at the 87.8th percentile. The discrepancy is due to the relatively small sample size. But we would have to conclude that there is relatively weak evidence against the hypothesis of homogeneity.

It is apparent when examining the data in detail that subjects with much infiltration are more likely to show an improvement than subjects with little infiltration. This suggests that the *degree* of improvement may differ. One way of examining this is to score the responses and compare the mean scores. One scoring scheme is to grade the responses $3, 2, 1, 0, -1$ corresponding to marked improvement through worsened levels of infiltration. The mean scores for the two groups are then of the form

$$f_j = 3(p_{1j}) + 2(p_{2j}) + 1(p_{3j}) + 0(p_{4j}) + (-1)(p_{5j}) \qquad j = 1, 2. \quad (3.19)$$

The null hypothesis of no difference in the degree of change is

$$H_0\colon \ f_1 = f_2 = f. \qquad (3.20)$$

To test this hypothesis it is required that we estimate f_j and the associated variances. The estimator of f_j is the usual mean based on grouped data:

$$\text{Little:} \quad \hat{f}_1 = 3(0.076) + 2(0.188) + 1(0.292) - (0.076) = 0.819$$

$$\text{Much:} \quad \hat{f}_2 = 3(0.135) + 2(0.288) + 1(0.308) - (0.019) = 1.269.$$

Next we need to estimate the associated variances. Since f_j are simply means, the usual sample estimators of variance can be used:

$$V(\hat{f}_j) = V_j = \frac{\Sigma_{i=1}\ a_i^2 \hat{p}_{ij} - \hat{f}_j^2}{n}$$

where a_i is score for the ith category and n rather than $n - 1$ is used to be consistent with maximum likelihood estimation. Thus we have

$$144 v_i = [9(0.076) + 4(0.188) + (0.292) + (0.076)] - (0.819)^2$$

$$= 1.804 - 0.671 = 1.133,$$

or $v_i = 1.133/144 = 0.0079$. Similarly, $v_2 = 0.0208$. We now have two choices. If the Pearson chi-square is to be used, then v_i and v_2 are used to form a pooled variance estimate. This approach is used by Snedecor and Cochran (1967, p. 245) to generate an approximate t-test. An alternative is the Neyman chi-square in the form of a *Wald statistic*: $Q = (0.819 - 1.269)^2/(0.0079 + 0.0208) = 7.06$. Since (3.20) corresponds to a single hypothesis ($f_1 - f_2 = 0$), there is only one degree of freedom, and the hypothesis is rejected. That is, Q is so large that we must conclude that subjects with much infiltration show a greater degree of improvement than those with little infiltration.

This example raises a number of issues. First, we have the somewhat contradictory result of accepting the hypothesis of homogeneity and rejecting the hypothesis of no difference in degree of improvement. This will occur not infrequently when one hypothesis is global with many degrees of freedom and the other is specific with few degrees of freedom. The consensus is that the more specific hypothesis is more sensitive and therefore more appropriate.

The second issue concerns the choice of scales. In this case we assumed that the levels of improvement were equally spaced. It can be shown that the Neyman chi-square will not change as long as this assumption is met regardless of the actual values of the scale. However, if the scale is not equally spaced, the statistic will be affected. The analogous problem for the Pearson chi-square is discussed by Snedecor and Cochran (1967, p. 246).

The question of alternative intervals for the scale can be addressed in part by comparing the one-degree-of-freedom test to the global test. Here some care must be exercised and the same estimators used in both test statistics. That is, the earlier computation of Q used maximum likelihood estimators, but weighted least squares estimators could have been used. The computations are a bit more involved but can be accomplished using standard software such as PROC FUNCAT (SAS 1982b). The required computer code is shown in Appendix 3A. The relevant results are shown in Table 3.7.

The first entry in Table 3.7 corresponds to the null hypothesis (3.20) of no difference in the degree of change. It is identical to the value that we computed manually. The third line, the total, corresponds to the hypothesis of homogeneity (3.18). Since weighted least squares rather than maximum likelihood estimates have been used, it is somewhat smaller than the earlier computation. Finally, the difference between the two is shown in the residual line. This corresponds to the test for nonlinearity. It is not only nonsignificant but is also less than

Table 3.7 Neyman Chi-Square Test Statistics for Leprosy Data

Source	H_0	df	Q	p-Value
Linear trend	(3.20)	1	7.06	0.01
Residual		3	1.42	0.70
Total	(3.18)	4	8.48	0.08

3.84, the 95th percentile of the chi-square distribution on one degree of freedom. This indicates that no single degree of freedom in the residual is significant. We can conclude that our choice of scales appears reasonable. Some alternative scales will be discussed in Chapter 4.

The final question we might pose about these data is: If the factor-response structure were reversed, could we test for a trend? Rather than thinking of change in health as a response, suppose that it defined five binomials where the response was degree of infiltration. Now we are posing the question as: Is the degree of infiltration correlated with the change in health? This can be answered descriptively by plotting the proportion with much infiltration as a function of change in health. This is shown in Figure 3.1. The estimated proportions, denoted q_i,

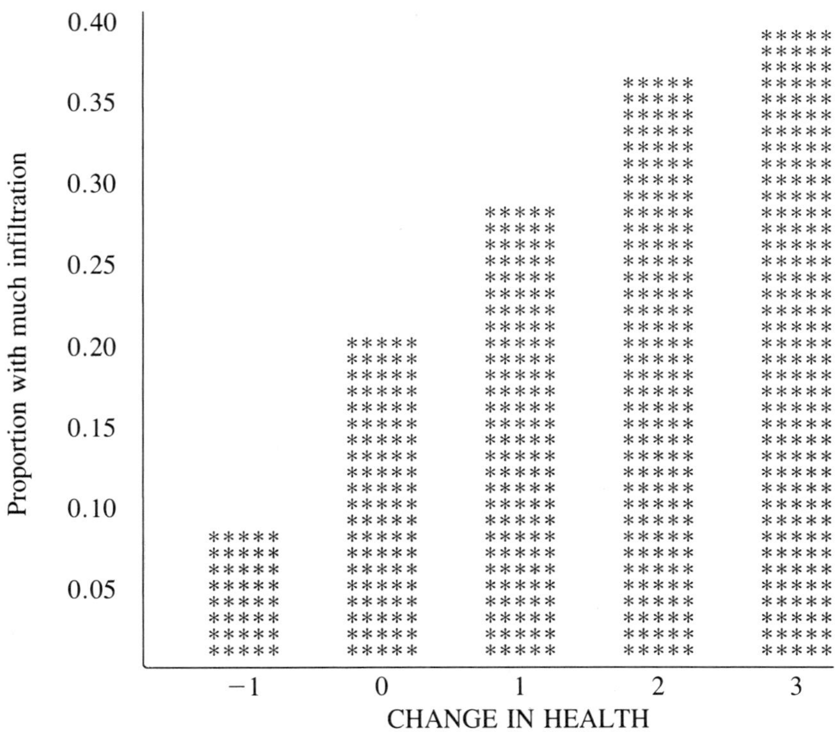

Figure 3.1 Proportion of leprosy patients with much infiltration as a function of change in health. (Courtesy of Snedecor and Cochran, 1967, p. 247.)

are shown in column (4) of Table 3.6. Clearly, there is marked evidence of an increasing trend as the change in health increases.

A statistical model that characterizes the trend is

$$E(q_i) = B_0 + B_1X_i. \tag{3.21}$$

Here B_0 and B_1 are the usual regression coefficients, X is the change in health score, and we have written the model in terms of the expected value of the estimators, q. This is similar to the earlier problem on infant mortality rates in New England (Section 2.8), with the distinction that the underlying distribution is assumed to be a binomial within each level of change. An ordinary least squares approach to the problem is found in Snedecor and Cochran (1967, pp. 246–248). We will use the weighted least squares estimators as an alternative for testing

$$H_0 \quad B_1 = 0. \tag{3.22}$$

The weighted least squares estimators are

$$b_1 = \frac{\Sigma_{i=1}^{I} W_i(X_i - \overline{X})(q_i - \bar{q})}{\Sigma_{i=1}^{I} W_i(X_i - \overline{X})^2}$$

$$b_0 = q - b_1X,$$

where

$$W_i = \frac{n_i}{q_i(1 - q_i)}$$

$$\overline{X} = \frac{\Sigma_{i=1}^{I} W_iX_i}{\Sigma_{i=1}^{I} W_i}$$

$$\bar{q} = \frac{\Sigma_{i=1}^{I} W_iq_i}{\Sigma_{i=1}^{I} W_i}$$

The corresponding Neyman chi-square is

$$Q = \sum_{i=1}^{I} W_i[q_i - (b_0 + b_1X_i)]^2. \qquad (3.23)$$

This tests that the model (3.21) is appropriate. The test for linear trend (3.22) can be obtained by computing the Neyman chi-square for homogeneity based on q and subtracting. Alternatively, the standard error of b_1 can be estimated and the appropriate z-test constructed.

We again use PROC FUNCAT (SAS 1982b) to perform the computations. The test statistics and estimators are shown in Figure 3.2. The levels of change correspond to the five samples and the response function is the proportion with much infiltration (A). Next, the test statistics are shown. The line labeled "change" gives the Neyman chi-square for (3.22). From this we conclude that there is a significant linear trend ($Q = 8.91$, df $= 1$, $p < 0.001$). The test for more complex trends is labeled "residual" and is clearly nonsignificant ($Q = 0.28$, df $= 3, p = 0.96$). The global test for homogeneity of the q_i is obtained by summing these Q statistics. In this case it is similar to the earlier test for homogeneity based on likelihood ratio statistics, but this is not generally true. Finally, the last two lines indicate the estimates and associated standard errors. Thus the estimated model corresponding to (3.21) is

$$\begin{aligned} q_i = \quad & 0.189 + \quad 0.080X_i, \\ & (\pm 0.035) \quad (\pm 0.027) \end{aligned}$$

where the q_i indicates predicted values and the terms in parentheses are the standard errors of the estimated model parameters.

Snedecor and Cochran (1967, p. 247) report the estimate of B_1 as 0.0756 with a standard error of 0.0293. This yields a z-test of 2.58, which corresponds to $X^2 = 6.66$. Since this test is based on ordinary least squares, the Neyman or weighted least squares approach can be shown to be superior (Neyman 1949). It is reassuring that the results are similar. A likelihood approach to the problem could also be undertaken, but this requires iterative procedures such as the Newton–Raphson method. This is discussed more fully in Haberman (1978), especially in Chapters 1 and 2.

SAMPLE	RESPONSE FUNCTION	DESIGN MATRIX 1	2
1	0.0833333	1	−1
2	0.19697	1	0
3	0.275862	1	1
4	0.357143	1	2
5	0.388889	1	3

SOURCE	DF	CHI–SQUARE	PROB
INTERCEPT	1	29.92	0.0001
CHANGE	1	8.91	0.0028
RESIDUAL	3	0.28	0.9642

EFFECT	PARAMETER	DF	ESTIMATE	CHI–SQ	PROB	STD
INTERCEPT	1	1	0.188739	29.92	0.0001	0.0345061
CHANGE	2	1	0.0800426	8.91	0.0028	0.0268105

Figure 3.2 Output from WLS analysis of leprosy data. (Courtesy of SAS Institute, 1982b, PROC FUNCAT).

3.4 $I \times J$ CONTINGENCY TABLES

The next level of complexity comes when both variables have more than two levels. As before, the hypotheses of independence and homogeneity yield the same chi-square test statistics, but the computations and interpretations are more complicated. The use of programmable calculators and packaged programs makes the computations relatively trivial, but the task of interpreting a "significant result" still rests with the analyst. In this section the test statistics and estimators are presented. We leave the question of summary indices to Section 3.5.

3.4.1 Hypotheses of Homogeneity

Jekel et al (1978) were interested in studying changing patterns in soft-tissue sarcomas over time. There are three principal types of these sarcomas: fibroid, lipoid, and mixed. The first group is characterized by tumors with a muscle cell origin. The lipoid tumors display a fat cell origin. The mixed group contains tumors with more specific pathological findings. The epidemiology of these cancers is relatively undeveloped, but there is little evidence to support a changing pattern over time. However, it is possible that diagnoses have become more specific, which could explain apparent increases in the incidence of the mixed type.

To study this question the investigators utilized data from the Connecticut Tumor Registry. The analysis was restricted to soft-tissue sarcomas of the arms and legs. After reviewing the detailed reports and eliminating questionable cases, they obtained 524 reports of soft-tissue sarcomas diagnosed between 1935 and 1974. It can be assumed that the cases were independently diagnosed, and also that within short periods, the chances of a given tissue pathology are relatively constant. These assumptions lead to viewing the data as a series of trinomials. If the dates of diagnosis are grouped into decades, this means that we wish to compare four such trinomials and test for homogeneity over time.

The tabulation of four trinomials will yield $4 \times 3 = 12$ frequencies to be considered. This is much more difficult to visualize than the previous contingency tables. One graphical aid is the *block chart*. An example is shown in Figure 3.3. The rows are the decades, the columns are the tissue types, and the heights indicate the frequencies. The frequencies are also printed in the cells. In the decades from 1935–

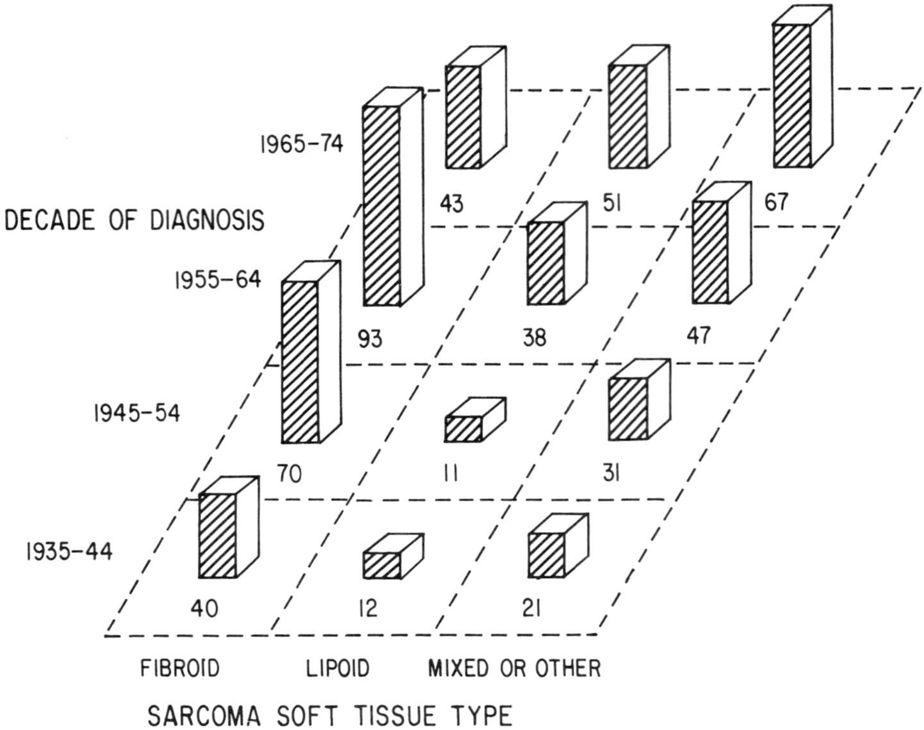

Figure 3.3 Block chart of tissue type by decade of diagnosis: vertical heights represent frequencies. (Courtesy of SAS Institute, 1982a.)

1944 to 1955–1964 the majority of sarcomas were of the fibroid type. But by the final decade this pattern changes dramatically. Another notable pattern is that while fibroid tumors peaked in 1955–1964, both the lipoid and mixed types increased systematically over the four decades. These general patterns need to be subjected to statistical analysis.

Let Y_{ij} denote the frequency of tumor type j (= 1, 2, 3) in decade i (= 1, 2, 3, 4). The expectation of Y_{ij} is given by $E(Y_{ij}) = m_{ij} = n_i p_{ij} > 0$, where $\Sigma_{j=1} p_{ij} = 1$ and n_i is the total tumors in the ith decade.

That is, let p_{ij} correspond to the probability of a tumor being classified as type j in decade i. The hypothesis of homogeneity is

$$H_0: \quad p_{ij} = p_{.j} \qquad j = 1, 2, 3. \tag{3.24}$$

To test the hypothesis, we need to estimate $p_{.j}$. The maximum likelihood estimators are

$$\hat{p}_{.j} = \frac{Y_{+j}}{n}. \tag{3.25}$$

So $\hat{p}_{.1} = (40 + 70 + 93 + 43)/524 = 246/524 = 0.470$. Similarly, $\hat{p}_{.2} = 0.214$ and $\hat{p}_{.3} = 0.317$.

The computation of degrees of freedom can be done three ways. Since there are three hypothesis, $H_0{}'$: $p_{1j} = p_{2j}, p_{1j} = p_{3j}, p_{1j} = p_{4j}$, and there are two parameters to be compared in each decade, we have $3 \times 2 = 6$ degrees of freedom. Generally, the number of hypotheses times the number of distinct parameters in each hypothesis yields the number of degrees of freedom.

Alternatively, the four trinomials imply four constraints of the form $\Sigma_{j=1}^{3} p_{ij} = 1$. To test (3.24), two values of $p_{.j}$ must be estimated. This leads to the computation df = cells − estimates − constraints = $12 - 2 - 4 = 6$. A final approach is simply to apply the rule df = $(I - 1)(J - 1) = (4 - 1)(3 - 1) = 6$.

With estimates and degrees of freedom in hand, we can compute the test statistics using

$$\hat{m}_{ij} = n_i \hat{p}_{ij}. \tag{3.26}$$

The results are shown as the second entries in Table 3.8. For example, in 1955–1964, 93 fibroid tumors were observed but only an estimated 83.6 would be expected if the diagnosis probabilities were homogeneous. The Pearson chi-square statistic is given by

$$X^2 = \frac{\Sigma_{i=1}^{4} \Sigma_{j=1}^{3} (y_{ij} - \hat{m}_{ij})^2}{\hat{m}_{ij}} \tag{3.27}$$

$$= \frac{(40 - 34.3)^2}{34.3} + \frac{(12 - 15.6)^2}{15.6} + \cdots + \frac{(67 - 51.0)^2}{51.0}$$

$$= 1.0 + 0.8 + \cdots 5.0 = 45.0.$$

Table 3.8 Limb Soft-Tissue Sarcomas by Decade of Diagnosis, Connecticut, 1935–1974 [y_{ij}, m_{ij}, $(y_{ij} - m_{ij})^2/m_{ij}$, and percentages]

	Tissue Type			
Decade	Fibroid	Lipoid	Mixed or Other	Total
1935–44	40	12	21	73
	34.3	15.6	23.1	
	1.0	0.8	0.2	
	7.63	2.29	4.01	13.93
	54.79	16.44	28.77	100.00
1945–54	70	11	31	112
	52.6	23.9	35.5	
	5.8	7.0	0.6	
	13.36	2.10	5.92	21.37
	62.50	9.82	27.68	100.00
1955–64	93	38	47	178
	83.6	38.0	56.4	
	1.1	0.0	1.6	
	17.75	7.25	8.97	33.97
	52.25	21.35	26.40	100.00
1965–74	43	51	67	161
	75.6	34.4	51.0	
	14.0	8.0	5.0	
	8.21	9.73	12.79	30.73
	26.71	31.68	41.61	100.00
Total	246	112	166	524
	46.95	21.37	31.68	100.00

Pearson chi-square	45.0	df = 6	prob = 0.0001
Likelihood ratio chi-square	47.5	df = 6	prob = 0.0001
Neyman chi-square	59.0	df = 6	prob = 0.0001

The Neyman chi-square statistic is computed similarly:

$$Q = \frac{\sum_{i=1}^{4} \sum_{j=1}^{3} (y_{ij} - \hat{m}_{ij})^2}{y_{ij}} \tag{3.28}$$

$$= \frac{(40 - 34.3)^2}{40} + \frac{(12 - 15.6)^2}{12} + \cdots + \frac{(67 - 51.0)^2}{67}$$

$$=0.8 + 1.1 + \cdots 3.8 = 59.0.$$

Finally, the likelihood ratio chi-square statistic is

$$G^2 = 2\sum_{i=1}^{4} \sum_{j=1}^{3} y_{ij} \ln\left(\frac{y_{ij}}{\hat{m}_{ij}}\right) \tag{3.29}$$

$$= 2\left[40 \ln\left(\frac{40}{34.3}\right) + 12 \ln\left(\frac{12}{15.6}\right) + \cdots + 67 \ln\left(\frac{67}{51.0}\right)\right]$$

$$= 12.30 + (-6.3) + \cdots + 36.6 = 47.5.$$

Clearly, all of these are highly significant when compared to the chi-square distribution with six degrees of freedom. We must conclude that tissue typing was not uniform across decades, but can we be more specific about the result?

3.4.2 Decomposition of Chi-Square

One approach is to inspect the cell-specific ''contributions to chi-square'' statistics. These are of the form $X_{ij}^2 = (y_{ij} - \hat{m}_{ij})^2/\hat{m}_{ij}$. The contributions to chi-square are the third entry in the cells of Table 3.8. If we use a one-degree-of-freedom criterion and 5 percent significance level, the large cells will exceed 3.84. On this basis of the contributions to chi-square, the number of fibroid and lipoid tumors in 1945–1954 and all tumors in 1965–1974 differs sharply from what would be expected from a homogeneous population.

A second approach would be to examine the ''standardized residuals.'' These are defined as $z_{ij} = (y_{ij} - \hat{m}_{ij})/\sqrt{\hat{m}_{ij}}$. Again using a 5 percent criterion we search for cells exceeding 1.96. Now we find that there are too many fibroid tumors and too few lipoid tumors in 1945–1954. Similarly, there are too few fibroid tumors and too many lipoid and mixed tumors in 1965–1974. This technique is discussed in greater detail in Haberman (1978). An important qualification of this approach is that many comparisons are possible, so some type of Bonferroni adjustment (Section 2.7) probably should be applied. Another limitation is that it does not lead to simple summaries of the data.

A third approach is known as decomposition of chi-square. The basic idea is to identify subtables that break up the chi-square statistic into more interpretable pieces. Maxwell (1961) and Iverson (1979) give detailed presentations of the technique. There are a number of ways of approaching the problem, and alternative decompositions may lead to different conclusions. The decompositions work best with the likelihood ratio chi-squares and should obey the following rules:

1. The number of subtables cannot be greater than the number of degrees of freedom in the original table.
2. Each cell frequency of the original table must appear as a cell frequency in one and only one subtable.
3. Each marginal total of the original table must appear as a marginal total of one and only one subtable.
4. Subtable cell frequencies not appearing in the original table must appear as marginal totals in a different subtable. Marginal totals not appearing in the original must appear as either cells or grand totals.

Obviously, applying these rules takes a little practice and experience.

In our example the most dramatic effect in the block chart was the decline in fibroid tumors in the last decade (Figure 3.3). The corresponding subtable is shown in Table 3.9A. Here the periods 1935–1964 and nonfibroid tumors have been condensed. The resulting $G^2 = 39.476$, df = 1, which is significant. The complementary table is shown in Table 3.9B. Here $G^2 = 4.250$, df = 2, which is not significant. Following rule 2, two more tables can be constructed. First, Table 3.9C dichotomizes the decades into 1965–1974 or all other decades combined and compares lipoid to mixed or other tumors. This yields $G^2 = 0.732$. Second, the table for fibroid versus all other tumors compared across the three decades from 1935 to 1964 (Table 3.9D) yields a G^2 of 3.001, df = 2. The decomposition can be summarized as in Table 3.10. Notice that the components sum exactly to total. As noted by Snedecor and Cochran (1967, p. 253), this will only

Table 3.9A Dichotomized Decade by Tissue

Decade	Tissue Type: All Other	Fibroid	Total
1935–64	160	203	363
	192.6	170.4	
1965–74	118	43	161
	85.4	75.6	
Total	278	246	524

Table 3.9B Lipoid and Mixed Sarcomas, 1935–1964

Decade	Tissue Type: Lipoid	Mixed or Other	Total
1935–44	12	21	33
	12.6	20.4	
1945–54	11	31	42
	16.0	26.0	
1955–64	38	47	85
	32.4	52.6	
Total	61	99	160

Table 3.9C Lipoid and Mixed Tissue by Dichotomized Decade

Decade	Tissue Type: Lipoid	Mixed or Other	Total
1935–64	61	99	160
	64.5	95.5	
1965–74	51	67	118
	47.5	70.5	
Total	112	166	278

Table 3.9D Fibroid Tumors by Decade, 1935–1964

Decade	Tissue Type		Total
	All Other	Fibroid	
1935–44	33	40	73
	32.2	40.8	
1945–54	42	70	112
	49.4	62.6	
1955–64	85	93	178
	78.5	99.5	
Total	160	203	363

be approximately true for the Pearson and Neyman statistics, which is why G^2 is preferred.

The reason the decomposition works is seen by expanding the G^2 statistics into a simpler computational formula (Iverson 1979):

$$G^2 = 2\sum_{i=1}^{I} \sum_{j=1}^{J} Y_{ij} \ln\left(\frac{Y_{ij}}{\hat{m}_{ij}}\right) = 2\sum_{i=1}^{I} \sum_{j=1}^{J} Y_{ij} \ln\left[(Y_{ij}) \frac{n}{n_{i+} Y_{+j}} \right]$$

$$= 2\left[\sum_{i=1}^{I} \sum_{j=1}^{J} Y_{ij} \ln (Y_{ij}) - \sum_{i=1}^{I} Y_{i+} \ln (n_{i+}) \right.$$

$$\left. = - \sum_{j=1}^{J} Y_{+j} \ln (Y_{+j}) + n \ln (n) \right].$$

Now rules 2 and 3 require that the subtables contain all the components of the original table with the appropriate sign. Rule 4 requires that any new cell frequencies be canceled out by also appearing as marginal totals.

It is now evident that only the (fibroid, all other) versus (1935–1964, 1965–1974) table is important. We can recombine the remaining tables to simplify our final summary. For example, Table 3.9B and 3.9D can be recombined to produce Table 3.11. The summary is

Table 3.10 Decomposition of Chi-Square (G^2) for Sarcoma Type by Decade

Table	Type	df	G^2
	Initial Decomposition		
3.7	Original	6	47.459
3.7A	Fibroid 1965–74	1	39.476
3.9B	Complement	2	4.250
3.7C	All other, 1964–74	1	0.732
3.9D	Fibroid, 1935–64	2	3.001
	Final Decomposition		
3.7	Original	6	47.459
3.7A	Fibroid, 1965–74	1	39.476
3.7C	All other, 1965–74	1	0.732
3.11	1935–64	4	7.251

Table 3.11 Soft-Tissue Sarcomas by Decade of Diagnosis, 1935–1964

Decade	Fibroid	Lipoid	Mixed or Other	Total
1935–44	40	12	21	73
	40.8	12.3	19.9	
1945–54	70	11	31	112
	62.6	18.8	30.5	
1955–64	93	38	47	178
	99.5	29.9	48.5	
Total	203	61	99	363

Pearson chi-square	6.872	df = 4	prob = 0.1428
Likelihood ratio chi-square	7.251	df = 4	prob = 0.1232

shown in the lower panel of Table 3.10. We can now state our conclusions:

1. Table 3.11 shows that the tumor-type distribution is homogeneous from 1935–1964.
2. Table 3.9 shows the lipoid versus mixed and other distribution was homogeneous between 1935–1964 and 1965–1974.
3. Table 3.9A shows that fibroid tumors were significantly less likely to be diagnosed in the final decade.

These conclusions depend, of course, on the validity of our original assumptions. Moreover, the decomposition procedure is somewhat ad hoc. The latter qualification does not affect our basic conclusion that the tissue-type distribution has changed over time (Table 3.9).

3.4.3 Independence and Quasi-independence

The analysis in Section 3.4.2 focused on whether a set of responses (tissue type) was homogeneous across the levels of a factor (decade). For cross-sectional study designs, such as sample surveys, the hypotheses are often formulated in terms of the independence of two responses. The usual assumptions lead to a single multinomial with $I \times J$ levels. That is, let Y_{ij} be the count in the (i, j) cell for a sample of size n. The multinomial parameters are p_{ij}, which are the probabilities of a randomly selected individual being at level i of the first response and level j of the second response. Under this model there is only one constraint, which can be written two ways:

$$(1)\quad \sum_{i=1}^{I} \sum_{j=1}^{J} p_{ij} = 1 \quad \text{or} \quad (2)\quad \sum_{i=1}^{I} \sum_{j=1}^{J} Y_{ij} = n. \tag{3.30}$$

The hypothesis of independence is

$$H_0\colon\ p_{ij} = p_{i+}p_{+j} \qquad i = 1, 2, \cdots, I;\ \ j = 1, 2, \cdots, J. \tag{3.31}$$

To test the hypothesis, we need to estimate the expected counts and apply the usual test of fit statistics.

The estimated expected counts are based on the maximum likelihood estimates under the null hypothesis:

$$\hat{p}_{i+} = \frac{Y_{i+}}{n} \qquad \hat{p}_{+j} = \frac{Y_{+j}}{n}. \tag{3.32}$$

This leads to

$$\hat{m}_{ij} = n\hat{p}_{i+}\hat{p}_{+j} = \frac{Y_{i+}Y_{+j}}{n}. \tag{3.33}$$

For example, in Table 3.9 the expected number of lipoid sarcomas in 1945–1954 is estimated to be $\hat{m}_{22} = 112(112)/524 = 23.9$. It should be evident that the expected counts under the hypothesis of independence (3.31) are identical to those under that of homogeneity (3.24), so the test statistics are identical. The degrees of freedom are also the same since we have 12 cells, 5 parameters, and 1 constraint: $12 - 5 - 1 = 6$; or in general, $IJ - (I - 1) - (J - 1) - 1 = (I - 1)(J - 1)$.

An interesting extension of this hypothesis is whether the responses are independent for most but not all of the cells. That is, suppose that we excluded fibroid tumors in 1965–1974. Does the table now correspond to a hypothesis of independence? Since we nearly have independence, this hypothesis is termed *quasi-independence*. The form is

$$H_0\colon \quad p_{ij} = p_{i+}p_{+j} \textit{ except for the excluded cells.} \tag{3.34}$$

For simple patterns such as excluding a few contiguous cells the estimation of m_{ij} is straightforward.

The table is first rewritten as in Table 3.12 where the marginal counts (Y_{i+}, Y_{+j}) are reduced by the amount of the excluded cell. Thus we now have a sample of 481 observations with 118 from 1965–1974 and 203 fibroid tumors. In the first step the expected values for the cells adjacent to the excluded cell are estimated based on the table which contains them. That is,

$$\begin{aligned}\hat{m}_{21} &= \frac{203(178)}{40 + 12 + 21 + 70 + 11 + 31 + 93 + 38 + 47} = \frac{6134}{363} \\ &= 9.543 \\ \hat{m}_{31} &= \frac{112(118)}{12 + 21 + 11 + 31 + 38 + 47 + 51 + 67} = \frac{13{,}216}{278} \\ &= 47.540.\end{aligned}$$

Table 3.12 General Block Triangular Table with I Rows and J Columns[a]

	Response 2		
Response 1	$1\ 2 \cdots J_1$	$J_1 + 1 \cdots J_1 + J_2$	$J_1 + J_2 + 1 \cdots J_1 + J_2 + J_3$
1 2 . . . I_1			"Diagonal"
$I_1 + 1$ $I_1 + 2$ . . . $I_1 + I_2$		"Diagonal"	Excluded
$I_1 + I_2 + 1$ . . . $I_1 + I_2 + I_3$	"Diagonal"	Excluded	Excluded

[a] $I = I_1 + I_2 + I_3$; $J = J_1 + J_2 + J_3$.
Source: Bishop et al. (1975, Fig. 2.5–5).

Next the marginal totals are adjusted to reflect these estimates. The results are shown in total row (1) and total column (1). For example, we now have 203 − 99.543 = 103.457 fibroid tumors to allocate to the decades before 1955. Notice that since the total for 1965–1974 must be 118, the estimate for mixed or other tumors is obtained by subtraction and the corresponding step 1 total is reduced to 0. The second step is to treat the estimated expected values as excluded and repeat step 1. That is,

$$\hat{m}_{21} = \frac{103.457(112)}{40 + 12 + 31 + 70 + 11 + 31} = \frac{11{,}587.184}{185} = 62.633$$

$$\hat{m}_{32} = \frac{64.460(78.457)}{12 + 21 + 11 + 31 + 38 + 47} = 31.608.$$

The marginal totals are again readjusted and the process continued until all of the expected values are estimated as in Table 3.13. Now the

Table 3.13 Estimated Counts Under Model of Quasi-independence

	Tissue Type					
	Fibroid	Lipoid	Mixed or Other	Total		
Decade	(1)	(2)	(3)	0	1	2
1935–44 (1)	40	12	21	73	73	32.176
	40.824	12.963	19.215			
1945–54 (2)	70	11	31	112	112	49.367
	62.633	19.889	29.481			
1955–64 (3)	93	38	47	178	78.457	0
	99.543	31.608	46.847			
1965–74 (4)	—	51	67	118	0	
		47.540	70.460			
Total						
0	203	112	166	481		
1	103.457	64.460	95.540		263.457	
2	0	32.852	48.696			81.543

$X^2 = 7.32$ $\quad Q = 10.23$ $\quad G^2 = 7.98$

$p = 0.20$ $\quad p = 0.07$ $\quad p = 0.16$

$\text{df} = 11 - 5 - 1 = 5$

test statistics for the fit of the data to the model are applied the usual manner.

The resulting chi-squares are shown at the bottom of Table 3.13 The degrees of freedom must also be adjusted since we only have 11 cells: $11 - 5 - 1 = 5$. Alternatively, we can view p_{41} as an extra parameter that has been estimated, yielding $12 - 6 - 1 = 5$. The Pearson and likelihood ratio statistics are small enough to suggest accepting the hypothesis of quasi-independence, but the Neyman chi-square is large enough to be somewhat circumspect in drawing unqualified conclusions. An additional point is to note that the G^2 statistic is identical to what is found through decomposition of the chi-square procedure in Tables 3.9B–3.9D. This suggests that the test of quasi-independence and the decomposition of chi-square technique can be used together to arrive at simple summaries of the data. One possibility is to search for significant subtables using the former. Then other subtables are constructed to complete the decomposition of chi-square.

The procedure just described can be applied to any table with a *block-triangular* structure. The general form of such tables is displayed in Table 3.12. At each step expected counts in cells adjacent to the excluded cells are estimated using

$$\hat{m}_{ij}^{*} = \frac{Y_{i+}^{*} Y_{+j}^{*}}{Y_{++}^{*}}, \tag{3.35}$$

where the Y_{i+}^{*}, Y_{+j}^{*}, and Y_{++}^{*} reflect adjustments from preceding steps. In each successive step only terms along the ''diagonal'' are adjusted. This formulation is based on the results reported in Bishop et al. (1975, pp. 194–199), to which the reader is referred for a more extended discussion.

Needless to say, if several cells are candidates for successive exclusion, the computations can be formidable. Moreover, if the candidates for exclusion do not exhibit a simple pattern, the estimation must be done iteratively. This suggests the procedure generally requires at least a microcomputer to be practical. The program BMDP4F in the BMDP statistical software package (Dixon 1981) is one commonly available program. The code for analyzing Table 3.8 is shown in Figure 3A.2.

3.5 INDICES OF ASSOCIATION

After finding that responses vary across the levels of a factor or that two response variables are dependent, analysts need to summarize their findings in a compact manner which can be easily communicated and interpreted. The most common method is to display the contingency table and associated p-value. This certainly provides a reviewer sufficient information to verify statistical conclusions but is somewhat limited. First, it is difficult to compare tables across studies with differing sample sizes and marginal distributions. Second, for tables more complex than 2×2, detailed inspection of patterns of association is cumbersome. Third, the p-value only tells the reader of the presence of an association, but not its strength or direction. For these reasons, and others, a variety of indices of association have been developed for summarizing findings. Detailed discussion of these measures are found in the monographs by Bishop et al. (1975) and Goodman and Kruskal (1979). The latter is a collection of four fundamental papers.

We will consider the problem for 2×2 tables and $I \times J$ tables. The former will allow us to examine the two approaches to association: correlation coefficients versus cross-product or odds ratios. This will also emphasize desirable properties of a measure. The examination of the larger table will demonstrate the difficulty of developing a single optimal measure.

3.5.1 2 × 2 Tables: Correlation and the Odds Ratio

Let the parameters of the 2×2 table be as shown in Table 3.14 with the respective random variables, Y_{ij}. The upper panel displays the data as a two-response study. The one-factor, one-response study can be generated as follows. Suppose that variable 1 is the factor; let $q_1 = p_{i1}/p_{i+}$. The result is shown in the lower panel of Table 3.14. The remaining connections between the two types of studies are made by noting $n_i = Y_{i+}$. Three indices of association for 2×2 tables have already been introduced:

(1) Simple difference: $$D = q_1 - q_2 = \frac{p_{11}}{p_{1+}} - \frac{p_{21}}{p_{2+}} = \frac{p_{11}p_{22} - p_{12}p_{21}}{p_{1+}p_{2+}} \tag{3.36}$$

(2) Relative risk: $$\text{RR} = \frac{q_1}{q_2} = \frac{p_{11}p_{2+}}{p_{21}p_{1+}} \tag{3.37}$$

(3) Odds ratio: $$\text{OR} = \frac{q_1(1 - q_2)}{q_2(1 - q_1)} = \frac{p_{11}p_{22}}{p_{21}p_{12}}. \tag{3.38}$$

One desirable property for an index of association is that it lie between -1 and $+1$, where $+1$ indicates perfect association and 0 indicates no association. Both the simple difference and odds ratios have simple transformations that yield this property or *norming*.

Table 3.14 Random Variables and Parameters of a 2 × 2 Table

Variable 1 (i)	Variable 2 (j) 1	2	Total
	Two-Response Study		
1	Y_{11}	Y_{12}	Y_{1+}
	p_{11}	p_{12}	p_{1+}
2	Y_{21}	Y_{22}	Y_{2+}
	p_{21}	p_{22}	p_{2+}
Total	Y_{+1}	Y_{+2}	n
	p_{+1}	p_{+2}	1.00
	1-Factor, 1-Response Study[a]		
1	Y_1	$n_1 - Y_1$	n_1
	q_1	$1 - q_1$	1.00
2	Y_2	$n_2 - Y_2$	n_2
	q_2	$1 - q_2$	1.00
Total	Y	$n - Y$	n
	q	$1 - q$	1.00

[a] $q = (n_1q_1 + n_2q_2)/n$.

The simple difference can be written in terms of the Pearson chi-square:

$$X^2 = \frac{n(p_{11}p_{22} - p_{21}p_{12})^2}{p_{1+}p_{2+}p_{+1}p_{+2}} = \frac{nD^2(p_{1+}p_{2+})}{p_{+1}p_{+2}} . \tag{3.39}$$

When the margins are symmetric (i.e., $p_{1+} = p_{+1}$) the maximum of X^2 is nD^2. Hence the normed simple difference is the *phi coefficient*:

$$\varphi = \sqrt{\frac{X^2}{n}} = D\sqrt{\frac{p_{1+}p_{2+}}{p_{+1}p_{+2}}} = (q_1 - q_2)\sqrt{\frac{n_1n_2}{y(n-y)}}. \tag{3.40}$$

For the two-response study φ has the particularly nice characteristic of being equal to the Pearson product moment correlation when the responses are scored 0, 1. For a derivation of this result, see Bishop et al. (1975, pp. 380–381). This means that it is the measure of choice when one undertakes a factor analysis with many dichotomous measures.

One possible estimator for the variance of φ can be obtained when we consider Y as fixed at y. Then only q_1 and q_2 are random so

$$V(\varphi) = V\left[(q_1 - q_2)\sqrt{\frac{n_1 n_2}{y(n-y)}}\right] = \frac{V(q_1 - q_2)n_1 n_2}{y(n-y)} \tag{3.41}$$

$$= \frac{n_2 q_1(1 - q_1) + n_1 q_2(1 - q_2)}{y(n-y)}.$$

An unconditional estimator is found in Bishop et al. (1975, p. 381).

To illustrate the phi coefficient, consider the data in Table 3.1. Here $X^2 = 34.428$ so $\varphi = \sqrt{34.428/2013} = 0.131$. Alternatively,

$$\varphi = (q_1 - q_2)\sqrt{\frac{n_1 n_2}{y(n-y)}} = \left(\frac{97}{297} - \frac{307}{1716}\right)\sqrt{\frac{297(1716)}{404(1609)}}$$

$$= 0.145(0.885) = 0.131.$$

The conditional variance of phi is

$$V(\varphi) = \frac{1716(97/297)(200/297) + 97(307/1716)(1409/1716)}{404(1609)}$$

$$= \frac{1716(0.327)(0.673) + 97(0.179)(0.821)}{650{,}036} = 0.000648.$$

So a 95 percent confidence interval is $0.131 \pm 1.96(0.025) = (0.08, 0.181)$. It should be evident now that although persons reporting that they work under tension are almost twice as likely to suffer a CHD event, the association is not terribly strong, although it is significant in a statistical sense.

A limitation of the phi coefficient is that its upper bound is determined by a sample's marginal distribution. This is evident when we consider the normed simple difference (3.40). Thus for a given level of association differing sample sizes n_1 and n_2 can yield different values of phi. The effect can be mitigated in part by drawing samples of equal size, but with observational data this is not always feasible.

An alternative approach is to rely on a measure that is free of this restriction. One such measure is the odds ratio (3.38). Suppose that samples of sizes $n_i^* = a_i n_i$ have been selected. The corresponding expected values are $m_i^* = n_i^* q_i = a_i n_i q_i = a_i m_i$. The resulting odds ratio is

$$\mathrm{OR}^* = \frac{m_1^*(n_2^* - m_2^*)}{m_2^*(n_1^* - m_1^*)} = \frac{a_1 m_1(a_2 n_2 - a_2 m_2)}{a_2 m_2(a_1 n_1 - a_1 m_1)} = \mathrm{OR}. \tag{3.42}$$

In words, the odds ratio is invariant under alternative sampling schemes, as noted by many authors.

The normed property led Yule (1900) to suggest

$$Q = \frac{\mathrm{OR} - 1}{\mathrm{OR} + 1} = \frac{p_{11}p_{22} - p_{12}p_{21}}{p_{11}p_{22} + p_{12}p_{21}}. \tag{3.43}$$

This is termed *Yule's Q* as distinct from the Neyman chi-square. It is evident that $-1 \leq Q \leq 1$. In addition, a little manipulation leads to the conclusion $Q = 0$ under independence or homogeneity.

An estimator of Q is

$$Q = \frac{Y_{11}Y_{22} - Y_{12}Y_{21}}{Y_{11}Y_{22} + Y_{12}Y_{21}}. \tag{3.44}$$

Based on the large-sample variance for the odds ratio, it can be shown (Bishop et al. 1975, p. 378) that a large-sample variance estimator is

$$V(Q) = \frac{(1 - Q)^2(1/Y_{11} + 1/Y_{12} + 1/Y_{21} + 1/Y_{22})}{4}. \tag{3.45}$$

We can apply this measure to the data in Table 3.1:

$$Q = \frac{97(1409) - 07(200)}{97(1409) + 307(200)} = \frac{75{,}273}{198{,}073} = 0.380$$

$$V(Q) = \frac{(1 - 0.380)^2(1/97 + 1/200 + 1/307 + 1/1409)}{4}$$

$$= 0.183(0.0193) = 0.00353.$$

Therefore, the 95 percent confidence interval is 0.380 ± 1.96(0.059) = (0.264, 0.496). This indicates substantially greater association than that based on phi, which reflects in part the maximum imposed by the marginal distribution. But primarily, it indicates a different approach to the concept of association.

Yule's Q does have the unpleasant property that if either $Y_{12} = 0$ or $Y_{21} = 0$, then $Q = 1$. Similarly, if $Y_{22} = 0$ or $Y_{11} = 0$, then $Q = -1$. Clearly, Q is overstating the degree of association in these circumstances.

3.5.2 The $I \times J$ Table: Nominal Variables

The 2 × 2 table provides some insight into indices of association, with phi being a correlational measure and Yule's Q being a measure that is unaffected by the marginal distributions. The next level of complexity is the $I \times J$ table. This introduces three issues. First, we generally want to preserve marginal invariance. Second, when one variable is a factor we would like to exploit this in the context of predictive explanations. Finally, when both variables are ordered, this should be incorporated into our analysis. We will not attempt to be exhaustive but restrict attention to measures that illustrate the main features of these approaches. In addition, the standard errors of the measures are computationally more awkward when we consider the $I \times J$ case. For a listing of more than two dozen measures and asymptotic standard errors the reader is referred to the BMDP manual (Dixon 1981, pp. 663–666). The measures we discuss are also available in SAS and SPSS (Nie et al. 1975) packages.

We begin our discussion with tables that are nominal on both variables. An example is the decade-by-tissue-type example (Table 3.8), where the ordered nature of decade is ignored. The simplest measure is the phi coefficient, where

$$\varphi = \sqrt{\frac{X^2}{n}} = \sqrt{\frac{45.004}{524}} = 0.293. \qquad (3.46)$$

Since the X^2 is invariant under permutations of the rows or columns, this property is preserved. Unfortunately, phi can exceed unity. One modification proposed by Pearson (1904) is the *contingency coefficient*:

$$C = \sqrt{\frac{X^2}{X^2 + n}} = \sqrt{\frac{45.004}{45.004 + 524}} = 0.281. \tag{3.47}$$

This has the limitation that $C < 1$. This was resolved by Cramer (1946), who proposed V:

$$V = \sqrt{\frac{X^2}{n \min(I - 1, J - 1)}} = \sqrt{\frac{45.004}{524(2)}} = 0.207. \tag{3.48}$$

Of these, V seems to be the most useful for comparing several tables of different dimensions.

The principal objection to these measures is that they are no more than index numbers. That is, they lack underlying substantive meaning. To provide a more interpretable measure, Goodman and Kruskal (1979) proposed the *proportional reduction in error* (PRE) *criterion* for indices of association, that is, how much less error prone is guessing the response when the factor is also known. For example, if we had to guess the tissue type for a randomly selected sarcoma in Table 3.8, our best guess would be fibroid since it occurred most often. Now if we also know the decade, then from 1935 to 1964 we would still guess fibroid. But in 1965–1974 the best guess would be mixed or other. The concept led Goodman and Kruskal to propose a measure based on the work of Guttman (1941):

$$L_{\text{RIF}} = \frac{\Sigma_{i=1} Y_{i\max} - Y_{\max}}{n - Y_{\max}}, \tag{3.49}$$

where $Y_{i\max}$ is the largest count in row i and $Y_{\max}$ is the largest column total. For example, the modal row counts are 40, 70, 93, and 67. The column total mode is 246. So $L_{\text{RIF}} = [(40 + 70 + 93 + 67) - 246]/(524 - 246) = 24/278 = 0.086$. Hence we do 8.6 percent better in guessing the tissue type when the decade is known.

The measure extends in a natural manner to the two-response case by considering the asymmetric lambdas: $L_{1|2} = (\Sigma_{i=1} Y_{i\max} - Y_{1\max})/(n - Y_{1\max})$, $L_{2|1} = (\Sigma_{j=1} Y_{\max j} - Y_{\max 2})/(n - Y_{\max 2})$. Then the symmetric lambda is given by

$$L_S = \frac{(\Sigma_{i=1} Y_{i\max} - Y_{1\max}) + (\Sigma_{j=1} Y_{\max j} - Y_{\max 2})}{2n - Y_{1\max} - Y_{\max 2}}. \quad (3.50)$$

For example, among the rows, $\Sigma_{i=1} y_{i\max} = 40 + 70 + 93 + 67 = 270$, $y_{1\max} = 246$. Among the columns, $\Sigma_{j=1} y_{\max j} = 93 + 51 + 67 = 211$, $y_{\max 2} = 178$. So the symmetric proportional reduction in error is

$$L_S = \frac{(270 - 246) + (211 - 178)}{2(524) - 246 - 178} = \frac{57}{624} = 0.091.$$

Thus by knowing the full table rather than the row and column tables, our error rate is reduced by 9.1 percent.

A parallel development for producing an interpretable measure of association is based on the likelihood ratio chi-square. Recall that G^2 can be written

$$G^2 = 2[\Sigma_{i=1}^{I} \Sigma_{j=1}^{J} Y_{ij} \ln (Y_{ij}) - \Sigma_{i=1}^{I} Y_{i+} \ln (Y_{i+}) - \Sigma_{j=1}^{J} Y_{+j} \ln (Y_{+j}) + n \ln (n)].$$

Theil (1972) and others have defined the *entropy* in a given cell as

$$U(i, j) = -\frac{Y_{ij}}{n} \ln \left(\frac{Y_{ij}}{n}\right). \quad (3.51)$$

Similarly, the marginal entropies are $U(i, .) = -(Y_{i+}/n) \ln (Y_{i+}/n)$ $U(., j) = -(Y_{+j}/n) \ln (Y_{+j}/n)$. The total uncertainty in the table is

$$U_S = \frac{2[\Sigma_{i=1}^{I} U(i, .) + \Sigma_{j=1}^{J} U(., j) - \Sigma_{i=1}^{I} \Sigma_{j=1}^{J} U(i, j)]}{\Sigma_{i=1} U(i, .) + \Sigma_{j=1} U(., j)}$$

$$= \frac{G^2}{n[\Sigma_{i=1}^{I} U(i, .) + \Sigma_{j=1}^{J} U(., j)]}$$

Thus the uncertainty coefficient, U_2, is analogous to phi but using the likelihood ratio chi-square. In the present example, $G^2 = 47.459$.

$$\sum_{i=1} U(i, .) = -\left[\left(\frac{73}{524}\right)\ln\left(\frac{73}{524}\right) + \left(\frac{112}{524}\right)\ln\left(\frac{112}{524}\right)\right.$$

$$\left. + \left(\frac{178}{524}\right)\ln\left(\frac{178}{524}\right) + \left(\frac{16}{524}\right)\ln\left(\frac{16}{524}\right)\right]$$

$$= 1.334$$

$$\sum_{j=1} U(.,j) = -\left[\left(\frac{264}{524}\right)\ln\left(\frac{246}{524}\right) + \left(\frac{112}{524}\right)\ln\left(\frac{112}{524}\right)\right.$$

$$\left. + \left(\frac{166}{524}\right)\ln\left(\frac{166}{524}\right)\right] = 1.049.$$

Hence $U_S = 47.459/(1.334 + 1.049)524 = 0.038$. Thus knowledge of the complete table reduces our uncertainty by 3.8 percent over knowledge of just the margins.

Asymmetric coefficients can be developed in a manner analogous to the asymmetric lambda. For example, consider the uncertainty coefficient for tissue type given decade. The general form is

$$U_{\text{RIF}} = \frac{G^2}{2n[\Sigma_{i=1} U(i, .)]} = \frac{47.459}{2(524)(1.334)} = 0.034. \tag{3.53}$$

So knowledge of decade reduces our uncertainty about tissue type by 3.4 percent.

The choices for nominal cross-classifications are among the following types of statistics: chi-square based, proportional reduction in error, and uncertainty measures. The first group does not have a ready interpretation and does not distinguish between the two study designs. The second group is appealing on those two counts but does have an important drawback. Namely, if the model value falls in only one column (or row), the measures are zero. The uncertainty coefficients are based on the likelihood ratio chi-square and do have an interpretation. Moreover, they depend on the entire distribution rather than the modal values. For these reasons, one may prefer them. The important point is that whenever one of these summary measures of association is reported, the analyst should think carefully about the choices and select whichever measures seem most appropriate to the particular study design.

3.5.3 $I \times J$ Tables: Ordinal Variables

In Sections 3.5.1 and 3.5.2, the indices of association require only that the data be cross-classified according to two nominal variables. Sometimes the levels of one or both variables have a meaningful ordering. For example, in Section 3.3 the degree of improvement is clearly ordinal and may correspond to an underlying continuum. Similarly, in Table 3.3, the variables age and frequency of breast self-examination are both ordered. This ordering of the variable levels may provide additional insight and suggest more sensitive indices of association. The case of one nominal variable and one ordered variable has been discussed, in part, in Section 3.3. We resume the discussion of alternative scales in Chapter 6. When both variables are ordered, correlation coefficients are an obvious possibility. If they also correspond to an underlying continuous variable and the categories are equally spaced, the Pearson product moment or Spearman rank correlation coefficients are natural indices of association. These are presented in most elementary texts and will not be considered further.

A special characteristic of categorical variables is that they need not correspond to an underlying continuum. This suggests that our measures should reflect underlying patterns rather than strict linearity as required for the usual correlation coefficients. By "pattern" we mean the concordance or discordance of the observation.

Suppose that a hypothetical table is defined by two variables A, the row variable, and B, the column variable. The levels of A are indexed by $i = 1, 2, \ldots, I$ and the levels of B are indexed by $j = 1, 2, \ldots, J$. By "ordering" we mean that if *level* i of A is less than or equal to *level* i', and *level* i' is less than or equal to *level* i'' of A, then *level* i must be less than or equal to *level* i'' of A.

Next, for concordance suppose that the indices of two observations are (i, j) and (i', j'). The observations are concordant if either $[(i < i')$ and $(j < j')]$ or $[(i > i')$ and $(j > j')]$. The observations are discordant if either $[(i < i')$ and $(j < j')]$ or $[(i > i')$ and $(j < j')]$. Let C be the number of concordant observations and D be the number of discordant observations. The remaining pairs are either tied on A $(i = i')$ or B $(j = j')$ or both. Let TA be the number of ties on A and TB the number of ties on B.

The simplest index of association based on patterns of concordance is gamma, suggested by Goodman and Kruskal (1979, p. 19):

$$\text{gamma} = \frac{C - D}{C + D}. \tag{3.54}$$

Gamma is defined unless the data are in a single row or column. It is +1 when the data are on the diagonal from left to right, and −1 when they are on the opposite diagonal. If A and B are independent, then gamma is zero. Finally, if $I = J = 2$, gamma equals Yule's Q. The large-sample variance of gamma is discussed by Goodman and Kruskal (1979, pp. 89–90 and 134).

To illustrate the computation of gamma, consider the data reported by Senie et al. (1981) on the relationship between age at diagnosis of breast cancer and the frequency of breast self-examination (Table 3.3). Let age and BSE be denoted by A and B respectively, with levels $i = 1, 2, 3$ and $j = 1, 2, 3$. The number of concordant pairs is obtained by multiplying the counts in a given cell by the concordant cell counts. Thus

$$C = 91(200 + 155 + 198 + 172) + 90(155 + 172)$$
$$+ 150(11{,}198 + 172) + 200(172) = 185{,}305$$

or in general,

$$C = \sum_{i=1}^{I} \sum_{j=1}^{J} Y_{ij}\left(\sum_{i'>i} \sum_{j'>j} Y_{i'j'}\right). \tag{3.55}$$

Similarly,

$$D = 90(150 + 109) + 51(150 + 200 + 109 + 198) + 200(109)$$
$$+ 155(109 + 198) = 126{,}202,$$

or in general,

$$D = \sum_{i=1}^{I-1} \sum_{j=2}^{J} Y_{ij}\left(\sum_{i'>i} \sum_{j'<j} Y_{i'j'}\right). \tag{3.56}$$

This yields

$$\text{gamma} = \frac{185{,}305 - 126{,}202}{185{,}305 + 126{,}202} = \frac{59{,}103}{311{,}507} = 0.19.$$

The corresponding large-sample standard error (using BMDP or SAS) is 0.038. Hence the test of H_0: gamma = 0 is based on $z = 0.190/0.038 = 5$, indicating a statistically significant concordance between age and reduced breast examination. The associated 95 percent confidence interval is $0.190 \pm 1.96(0.038) = (0.12, 0.26)$.

From a basis in nonparametric statistics, Kendall (1949) suggested an index of association that modifies gamma so as to adjust for the presence of ties. This is tau-b:

$$\text{tau-}b = \frac{C - D}{\sqrt{(C + D + \text{TA})(C + D + \text{TB})}}. \tag{3.57}$$

To illustrate, we first compute TA and TB:

$$\begin{aligned}\text{TA} &= 91(90 + 51) + 90(51) + 150(200 + 155) + 200(155) \\ &\quad + 109(198 + 72) + 198(172) = 176{,}057 \\ \text{TB} &= 91(150 + 109) + 150(109) + 90(200 + 198) + 200(198) \\ &\quad + 51(155 + 172) + (155 + 172) = 158{,}676.\end{aligned}$$

The general form is

$$\begin{aligned}\text{TA} &= \sum_{i=1}^{I-1} \sum_{j=1}^{J} Y_{ij}\left(\sum_{j'>j} Y_{ij'}\right) \\ \text{TB} &= \sum_{i=1}^{I} \sum_{j=1}^{J-1} Y_{ij}\left(\sum_{i'>i} Y_{i'j}\right).\end{aligned} \tag{3.58}$$

Using these, we obtain

$$\text{tau-}b = \frac{59{,}103}{\sqrt{(311{,}507 + 176{,}057)(311{,}507 + 158{,}676)}} = 0.123.$$

An alternative approach is useful when we think of one variable (e.g., age) as fixed and we wish to predict the other. This is known as *Somers'* $d_{B|A}$ (Somers 1962):

$$d_{B|A} = \frac{C - D}{C + D + \text{TA}} = \frac{59{,}103}{311{,}507 + 176{,}057} = 0.121. \tag{3.59}$$

The asymptotic standard error for both these measures is 0.025. The computational formulas are found in Kendall (1949) and Goodman and Kruskal (1979, p. 136).

There is no clear guidance as to which measure is preferred. Gamma has the advantage of simplicity. Moreover, it reduces to Yule's Q in the case of the 2×2 table; hence it fits into the earlier development of the odds ratio. The problem that it is 1 for patterns indicating less than perfect association still holds. For examples of these patterns, see Goodman and Kruskal (1979, p. 20). Finally, since it ignores ties it tends to overstate the degree of association. For this last reason many analysts prefer tau-b for the two-response study and Somers' d for the one-factor, one-response study.

APPENDIX 3A: SAS CODE FOR WLS ANALYSIS OF SCORED LEPROSY DATA

The data in Table 3.6 are input as a frequency table in lines 1–18 of Figure 3A.1. This is only one of many ways of inputting the data. The code for the procedure is in lines 19–23 and is discussed in the *SAS User's Guide: Statistics* (SAS 1982b). The reason for showing it here is to emphasize the structure of the problem. In line 20 the factor-response structure is defined by the model statement. Here the response, CHANGE, is set equal to the factor, INF. Optionally, we have requested the printing of the "design matrix," (X). The use of design

```
1   DATA;
2     INPUT CHANGE C1  C2;
3     ARRAY C  C1  C2;
4     DO OVER C;
5       COUNT  =  C;
6       INF  =  _I_;
7       OUTPUT;
8     END;
9     DROP  C1  C2;
10    LABEL CHANGE  =  CHANGE IN HEALTH
11            INF  =  DEGREE OF INFILTRATION
12            COUNT  =  CELL FREQUENCY
13    ; CARDS;
14    -1   11    1
15     0   53   13
16     1   42   16
17     2   27   15
18     3   11    7
19    PROC FUNCAT;
20      MODEL CHANGE  =  INF  /  X  ;
21      RESPONSE 1 0 0 0 0 / 0 1 0 0 0 / 0 0 1 0 0 / 0 0 0 1 0;
22      RESPONSE  -1  0  1  2  3  ;
23      WEIGHT COUNT;
```

Figure 3A.1 SAS computer code for weighted least squares analysis of leprosy data.

matrices will be discussed more extensively in Chapters 4 and 6. The first response statement, line 21, simply requests that only p_{1j}, p_{2j}, p_{3j}, and p_{4j} will be used in the analysis. The constraint $\Sigma_{i=1} p_{ij} = 1$ means p_{5j} need not be considered. The "/" indicates that each proportion be estimated separately. The second response statement, line 22, requests that the f_j be estimated as in (3.19). Finally, the weight statement, line 23, simply informs the program that input is in the form of a frequency table. The program computes the appropriate estimates, variances, covariances, and test statistics.

```
/PROBLEM     TITLE IS 'QUASI-INDEPENDENCE'.
/INPUT       VARIABLES ARE 2.
             TABLE IS 3,4.
             FORMAT IS '12F3.0'.
/VARIABLE    NAMES ARE TYPE, DECADE.
/TABLE       INDEX IS TYPE, DECADE.
             INITIAL IS 1.
/FIT         MODEL IS T,D.
             CELL IS STAN.
             STEP IS 6.
/PRINT       LAMB. STAND.
/END
40 12 21 70 11 31 93 38 47 43 51 67
```

Figure 3A.2 BMDP code for testing quasi-independence in Table 3.9.

4

Three-Way Tables

The investigation of whether an association between two variables changes when other variables are considered leads immediately to tables in three or more dimensions. In an example discussed in Chapter 1, the question of whether the effect of stepped-care treatment depends on the presence of end-organ damage is central to understanding the results of the Hypertension and Detection Follow-up Program. In this chapter we explore the analysis of three variables in some detail. Initially, the focus is on a single response variable (*dependent variable*) and two factors (*independent variables*). When the response has more than two levels and the levels are ordered, some alternative scoring schemes will be examined. Analyses of one response varying across the levels of two factors are based on inferences about homogeneity, similar to the hypotheses of homogeneity found in Chapter 3.

Other situations are more appropriately addressed in terms of the independence of three response variables. Here we are concerned about the associations among the three pairs of variables, while controlling or adjusting for the effect of a third. This introduces the possibility that an association depends on which level of the third variable is considered.

Finally, for some studies we may wish to reverse the factor-response structure. For example, in case-control studies the presence of disease is the factor and variables measured in the study are actually responses. However, using the odds ratio allows us to treat the disease as if it were a response and the measured variables as if they were factors.

For both models of independence and those where the factor-response structure should be reversed, the log-linear model is especially well suited. This very famous model is discussed at length in Sections 4.3 and 4.4. The chapter concludes by examining questions about model selection (Section 4.5), collapsibility (Section 4.6), and interpretation of results (Section 4.7). The conceptual structure generalizes to analyzing four or more variables, which is discussed in Chapters 5–7.

4.1 MODELS OF HOMOGENEITY

The problem of interaction among three categorical variables was first formulated by Bartlett (1935). The data he used to illustrate the problem are shown in Table 4.1. The top panel of the table shows the full three-dimensional table. For each combination of planting time (i = 1, 2: at once, spring) and length of cutting (j = 1, 2: long, short) the corresponding response distribution for survival (k = 1, 2: alive, dead) is shown. Among long cuttings planted at once, 65 percent or 156 of 240 survive. For the corresponding survival rate among short cuttings planted in the spring, only 31 of 240 or 12.9 percent survive. The "marginal totals" are shown at the bottom of Table 4.1. For example, when the data are collapsed over length of cutting, the marginal survival rate for planting at once is 54.8 percent, but for spring planting it is only 24 percent. Similarly, collapsing time of planting yields a 50 percent survival rate among long cuttings and a 28.8 percent survival rate among short cuttings.

The data array suggests an underlying experimental design where an equal number of plum root stalks have been randomized to each of four treatment groups. The treatment groups result from crossing the two levels of planting time with the two levels of length of cutting. This suggests four distinct questions about the survival rates:

Table 4.1 Bartlett's Data on Plum Root Stalks According to Length of Cutting and Time of Planting

Planting Time (i)	Length of Cutting (j)			Survival: Alive $(i, j, 1)$	Survival: Dead $(i, j, 2)$	Total $(i, j, +)$
At once	Long	$(1, 1, k)$	y_{11k}	156	84	240
			%	65.0	35.0	100.0
	Short	$(1, 2, k)$	y_{12k}	107	133	240
			%	44.6	55.4	100.0
Spring	Long	$(2, 1, k)$	y_{21k}	84	156	240
			%	35.0	65.0	100.0
	Short	$(2, 2, k)$	y_{22k}	31	209	240
			%	12.9	87.1	100.0
Total at once		$(i, +, k)$	y_{+1k}	263	217	480
			%	54.8	45.2	100.0
Total spring		$(2, +, k)$	y_{+2k}	115	365	480
			%	24.0	76.0	100.0
	Total long	$(+, 1, k)$	y_{1+k}	240	240	480
			%	50.0	50.0	100.0
	Total short	$(+, 2, k)$	y_{2+k}	138	342	480
			%	28.8	71.2	100.0
Total		$(+, +, k)$	y_{++k}	378	582	960
			%	39.4	60.6	100.0

Source: Bartlett (1935).

1. Is there statistically significant variation across the four groups?
2. Does planting time have a significant effect?
3. Does length of cutting have a significant effect?
4. Does the effect of one variable (factor) depend on the level of the other variable (factor)?

The first question is "global" in the sense that it includes the remaining questions. That is, if any of questions 2–4 are answered in the affirmative, the answer to question 1 is also yes. Conversely, if the answer to question 1 is no, the answer to the remaining questions is also no.

Data analysis begins by making some assumptions about the underlying experiment. Let Y_{ij1} denote the number of surviving plum root stalks at planting time i and cutting length j. For each treatment combination, assume that the sample size is fixed at $n_{ij} = 240$. If we assume that the outcomes for each stalk are independently and identically distributed, and there is a fixed probability of survival, q_{ij}, then Y_{ij1} follows a binomial distribution with parameters n_{ij} and q_{ij}, where n_{ij} is known and q_{ij} is to be estimated.

The global question 1 corresponds to the hypothesis

$$H_0\colon \quad q_{11} = q_{12} = q_{21} = q_{22} = q. \tag{4.1}$$

That is, we are hypothesizing constant probabilities of survival. We analyzed this problem in Section 3.3. It follows that the same test statistics can be used.

Also, notice that hypothesis (4.1) can be rewritten in terms of three distinct hypotheses:

$$H_0\colon \quad q_{11} - q_{12} = 0, \quad q_{11} - q_{21} = 0, \quad q_{11} - q_{22} = 0. \tag{4.2}$$

These three hypotheses correspond to the three degrees of freedom for the hypothesis test. Alternatively, there are four random variables $(Y_{111}, Y_{121}, Y_{211}, Y_{221})$ and one parameter (q) to be estimated, hence df $= 4 - 1 = 3$.

For the Pearson and likelihood ratio chi-square the maximum likelihood estimate (m.l.e.) of q is $\hat{q} = Y_{++1}/n = 378/960 = 0.3938$. The estimated expected counts are given by

$$\hat{m}_{ij1} = \hat{q}n_{ij} \qquad \text{and} \qquad \hat{m}_{ij2} = (1 - \hat{q})n_{ij}. \tag{4.3}$$

Since $n_{ij} = 240$ for all four treatments, $\hat{m}_{ij1} = 0.3938(240) = 94.5$ for all four expected survival counts. Similarly, $\hat{m}_{ij2} = 145.5$. Using these estimates in (2.46) and (2.47), we obtain $X^2 = 141.05$ and $G^2 = 151.02$.

The Neyman chi-square can be computed using the m.l.e.'s, but to illustrate the procedure for weighted least squares, the estimate is based on

$$\tilde{q} = \frac{\sum_{i=1}^{I} \sum_{j=1}^{J} v_{ij}^{-1}\, q_{ij}}{\sum_{i=1}^{I} \sum_{j=1}^{J} v_{ij}^{-1}} \tag{4.4}$$

where $v_{ij} = \hat{q}_{ij}(1 - \hat{q}_{ij})/n_{ij}$. In our example

$$\tilde{q} = \left[\frac{240(0.35)}{0.65(0.35)} + \cdots + \frac{240(0.129)}{0.129(0.871)}\right] \Big/ \left[\frac{240}{0.65(0.35)} + \cdots + \frac{240}{0.129(0.871)}\right]$$

or $\tilde{q} = 0.3382$. It follows that $\tilde{m}_{ij1} = 0.3382(240) = 81.17$ and $\tilde{m}_{ij2} = (1 - 0.3382)(240) = 158.83$. Using these values in (2.48), we obtain $Q = 207.19$.

Whether we use X^2, G^2, or Q, it is clear that the global hypothesis of homogeneity (4.1) should be rejected. This means that there are one or more significant differences among the treatment groups. The next step is to identify these differences. Our search for the differences will in turn answer questions 2, 3, and 4.

One way to search systematically for the statistically significant differences is to hypothesize and estimate a model for the q_{ij}. One possible model is

$$q_{ij} = M + P_i + L_j + PL_{ij}. \tag{4.5}$$

Here M is an overall mean, P_i characterizes the planting effect, L_j the length effect, and PL_{ij} the joint effect of the two.

The problem with this model is that it has too many parameters. Notice that there are at most four values of q_{ij}. But there are nine model parameters: $M, P_1, P_2, L_1, L_2, PL_{11}, PL_{12}, PL_{21}, PL_{22}$. When there are more model parameters than underlying probabilities, the model is called *overparameterized*. This is just like in basic algebra when we had four equations and nine unknowns: an infinite number of solutions are possible.

To solve this problem, an arbitrary set of constraints or *identifiability conditions* can be imposed to allow solving (4.5) in terms of the q_{ij}. Conditions similar to those often used in analysis of variance have the form

$$P_1 + P_2 = 0 \quad \text{or} \quad P_2 = -P_1 = -P$$

$$L_1 + L_2 = 0 \quad \text{or} \quad L_2 = -L_1 = -L$$

$$PL_{11} + PL_{12} = 0 \quad \text{or} \quad PL_{12} = -PL_{11} = -PL$$

$$PL_{11} + PL_{21} = 0 \quad \text{or} \quad PL_{21} = -PL_{11} = -PL$$

$$PL_{12} + PL_{22} = 0 \quad \text{or} \quad PL_{22} = -PL_{12} = PL_{11} = PL.$$

Using summation notation, this is the familiar

$$\sum_i^I P_i = 0 \qquad \sum_j^J L_j = 0 \qquad \sum_i^I PL_{ij} = \sum_j^J PL_{ij} = 0. \tag{4.6}$$

These conditions allow the model to be rewritten:

$$\begin{aligned} q_{11} &= M + P + L + PL, & q_{12} &= M + P - L - PL, \\ q_{21} &= M - P + L - PL, & q_{22} &= M - P - L + PL. \end{aligned} \tag{4.7}$$

For more complex problems this detailed presentation of a model is somewhat cumbersome. It can be written more compactly with the use of *indicator variables*. For example, let

$$X_1 = 1 \quad \text{when } i = 1, \ -1 \quad \text{when } i = 2;$$

$$X_2 = 1 \quad \text{when } j = 1, \ -1 \quad \text{when } j = 2. \tag{4.8}$$

Using these two indicator variables, the model (4.7) can be written as

$$q_{ij} = M + X_1P + X_2L + (X_1X_2)PL. \tag{4.9}$$

Indicator variables can be used to characterize the underlying experimental design so that they are really *design variables,* but since they usually take on values such as 1, 0, and -1, they are also termed *dummy variables.*

It should be emphasized that the choice of identifiability conditions (4.6) was quite arbitrary. Alternative conditions are used in different applications. This affects the specific parameter estimates, but under quite general conditions concerning the choice of indicator variables it does not affect the test statistics.

Given our identifiability conditions we can solve for the model parameters in terms of the underlying survival probabilities. That is the identifiability conditions lead to four equations (4.7) in four unknowns. Using elementary algebra, we obtain

$$M = \frac{q_{11} + q_{12} + q_{21} + q_{22}}{4} = \frac{\Sigma_i^2 \Sigma_j^2 q_{ij}}{4} = \frac{q_{++}}{4} \tag{4.10}$$

$$P = \frac{(q_{11} - q_{21}) + (q_{12} - q_{22})}{4} = \frac{q_{1+} - q_{2+}}{4} \tag{4.11}$$

$$L = \frac{(q_{11} - q_{12}) + (q_{21} - q_{22})}{4} = \frac{q_{+1} - q_{+2}}{4} \tag{4.12}$$

$$PL = \frac{(q_{11} - q_{12}) - (q_{21} - q_{22})}{4} \tag{4.13}$$

$$= \frac{(q_{11} - q_{21}) - (q_{12} - q_{22})}{4} \tag{4.14}$$

These solutions correspond nicely to the original interpretation of the parameters following the model definition (4.5). M corresponds to the average of the survival rates. P compares the survival rates for stalks planted at once (q_{1+}) to those for stalks planted in spring (q_{2+}). Similarly, L compares the rates for long stalks (q_{+1}) to those for short stalks (q_{+2}). For this reason, we say that P and L are the *main effects* of planting time and length of cutting, respectively.

Finally, the PL term can be viewed in two ways. First, it can be seen as a comparison of the length of cutting effect for stalks planted at once to the same effect for stalks planted in the spring. Second, it can be rewritten to be the comparison of the time of planting effect among long cuttings to the time effect among short cuttings. These two interpretations correspond to (4.13) and (4.14), respectively. Obviously, one interpretation implies the other. Because this term reflects the joint effect of the two treatments, it is usually termed the *interaction effect*.

The preceding discussion gives an interpretable model with which to characterize the variation in the q_{ij}. But since there are four model parameters and four q_{ij} the solution is exact, given any set of estimates $\hat{q}_{ij}$. For this reason it is referred to as a *saturated model*. Since such models do not simplify the interpretation, they are generally of little interest except as an initial characterization of the data. A more interesting model occurs when one or more terms have been set equal to zero. As a rule it is the most complex terms which are eliminated; this corresponds to the interaction term, PL.

The model with only main effects is

$$q_{ij} = M + X_1P + X_2L, \tag{4.15}$$

where the X_1 and X_2 dummy or design variables are still defined by (4.8). We are now prepared to estimate the parameters of the model with the hope that they will provide a simple interpretation of the variation in the q_{ij}. Weighted least squares is a simple computational procedure for estimating the model parameters. The form of the estimators is presented in Appendix 4A.1. It is enough to know that the procedure is entirely analogous to that found in Section 2.8. Moreover, computer software for generating weighted least squares estimates and the associated standard errors is readily available (SAS Institute 1982b: PROC FUNCAT). Maximum likelihood estimators can also be obtained for this model using the GLIM software package (Baker and Nelder 1978).

The first part of the analysis is termed the *analysis of variation*. It corresponds to breaking the global hypothesis (4.1) into pieces corresponding to the various main effects and the fit of the data to the model. The line labeled "total," (see p. 169) corresponds to the global hypothesis. The next line up, "residual," tests the hypothesis that the data fit the model. Since $Q = 0.08$ and df $= 1$, we can safely accept this hypothesis. This also answers the fourth question posed at the beginning of this section: No, the effect of one factor does not depend on the level of the other factor. Putting the issue differently, accepting the fit of the data to the model is equivalent to accepting the hypothesis

$$H_0\colon \quad PL = 0 \tag{4.16}$$

The next two lines up from the residual show the test statistics for the effects of time of planting and length of cutting. These correspond to

$$H_0\colon \quad P = 0 \qquad \text{and} \qquad H_0\colon \quad L = 0. \tag{4.17}$$

The corresponding Q statistics are quite large, indicating that both hypotheses should be rejected. The top line, "model," is the simultaneous test of

$$H_0\colon \quad P = L = 0. \tag{4.18}$$

Clearly, it can be rejected. Notice that the chi-squares for "model" and "residual" add exactly to "total," 207.11 + 0.08 = 207.19; but those for time and length do not add to the model. This illustrates the general result that although model and residual always partition the total, the model is not exactly decomposed by its components.

We have now answered the remaining two questions posed at the beginning of the section: (2) Yes, planting time has a significant effect; (3) Yes, length of cutting has a significant effect. Using the parameter estimates, the overall average survival is 39.3 percent after adjusting for time of planting and length of cutting. Planting at once increases this by 15.5 percentage points and planting in the spring decreases survival. The absolute difference is 31 points. Similarly, long cuttings have a 21.4-point advantage over short cuttings. The standard errors permit us to convert these point estimates into confidence intervals if that is desired.

Finally, using the estimated model parameters, we can generate predicted values. For example, for short cuttings planted at once, the predicted survival rate is 0.393 + 0.155 − 0.107 = 0.441. Notice that both in interpreting the parameters and generating the predicted values, we utilized the dummy variables (4.8) which characterized the study design. A different set of dummy variables would have affected our interpretation of the results, but for a broad range of possibilities, some of which will be illustrated later, the hypothesis tests would not be affected.

4.2 SCORED RESPONSES

In Section 4.1 we introduced the main ideas associated with models of homogeneity, that is, *two-factor, one-response* models. Although the factor structure and the associated design variables (dummy variables) are completely general, the dichotomous response tended to suggest fairly simple functions. In the example, the probability of survival was analyzed directly. Alternatively, the log-probability would have led to a relative risk model. When logistic regression is discussed (Chapter 7), a third simple response function, the log odds, will be considered. However, for many data sets, the response has more than two levels. When the categories are nominal the analysis is considerably more complex in that it is analogous to multivariate analysis of variance (MANOVA).

This situation is generally beyond the scope of this book. A introduction to the problem based on "log-linear models" is found in Haberman (1979, Chap. 6).

The response is often ordinal or may be based on an underlying continuum. An example of this was already considered in Section 3.3, where mean scores were compared across two populations. This type of analysis extends immediately to multiple-factor designs.

The mean score approach assumes that the spacing of the categories have an underlying substantive meaning. For example, if there are five categories, $j = 1, 2, 3, 4, 5$, then mean scores assume that a shift from category 1 to category 2 is the same as a shift from category 3 to category 4. This may be an unduly restrictive assumption.

Frequently, we will want to assume no more than that the categories are ordinal. In this case we are assuming only that if category 1 is less than category 2 and category 2 is less than category 3, then category 1 is less than category 3. In this situation ranks or midranks for the categories are a particularly appealing scoring system since it accounts for order without assuming continuity. A drawback is that rank scoring lacks ready interpretability. As an alternative, Bross (1958) suggested scoring the categories relative to some "defined population." These are termed *ridit scores*. It turns out that when the ridit is based on the marginal distribution, the ridit scores are equivalent to the rank scores (Selvin 1977). We will examine each of these approaches and apply them to a data set with two factors and a response with five levels.

The construction of rank scores is best illustrated with a simple example using continuous variables. West (Colton 1974, p. 113) reported the lengths of remission for patients suffering two types of depression: endogenous and neurotic. The number of days of remission between two successive attacks is shown for 12 patients of each type in Table 4.2. Here the data have been sorted in ascending order by length of remission. The null hypothesis is that the two groups have been sampled independently from the same population. That is, the labels "endogenous" or "neurotic" have been assigned to depressed patients at random, at least with respect to the length of remission.

The ranks are assigned to the combined sample of 24 patients in order of each patient's length of remission. Thus the patients with lengths 87, 91, and 101 receive ranks 1, 2, and 3, respectively. Three patients have length 105, which corresponds to ranks 4, 5, and 6. They

Table 4.2 Days of Remission and Rank for Patients with Endogenous or Neurotic Depression

Type	Length of Remission											
Endogenous												
Days	101,	105,	105,	109,	140,	140,	179,	214,	529,	744,	1547,	1818
Rank	3,	5,	5,	7,	8.5,	8.5,	10,	12,	15,	19,	23,	24
Neurotic												
Days	87,	91,	105,	199,	279,	479,	546,	602,	643,	794,	844,	1296
Rank	1,	2,	5,	11,	13,	14,	16,	17,	18,	20,	21,	22

Source: West (1969) reported by Colton (1974).

are all assigned the average of these three ranks, or 5. This average rank for tied observations is termed the *midrank*. The process of assigning ranks is continued until the last rank (24) is assigned to length 1818. Notice that the observations tied at 140 received $(8 + 9)/2 = 8.5$ as their midrank. These ranks can be used to conduct a Wilcoxon rank sum test for comparing the two groups. If several groups of equal size are to be compared, the Friedman chi-square can be employed. These tests are discussed in most elementary texts on nonparametric statistics (e.g., Lehman 1975). However, these tests are most appropriate when there are few ties. With categorical data the number of ties is generally quite large.

For the general problem, suppose that we have IJ samples each of size n_{ij}, $i = 1, \ldots, I$, and $j = 1, \ldots, J$. Assume that observations for each sample follow a multinomial distribution with K ordered categories, where the distribution parameters are $q_{ijk} > 0$, $\Sigma_{k=1}^{K} q_{ijk} = 1$ and $i = 1, \ldots, I$, $j = 1, \ldots, J$. Let the multinomial random variables be denoted Y_{ijk}. The midranks for the categories are defined to be

$$\begin{aligned}
R_1 &= \frac{Y_{++1} + 1}{2} \\
R_2 &= \frac{Y_{++1} + (Y_{++2} + 1)}{2} \\
&\;\;\vdots \\
R_k &= \frac{\Sigma_{k'=1}^{k-1} Y_{++k'} + Y_{++k} + 1}{2}.
\end{aligned} \tag{4.19}$$

Dividing these midranks by the sample size, n, yields the *rank scores*:

$$\bar{r}_k = \frac{\Sigma_{k'=1}^{k-1} Y_{++k} + (Y_{++k} + 1)/2}{n} = \frac{R_k}{n} \qquad k = 1, \ldots, K \tag{4.20}$$

where $n = \Sigma_{i=1}^{I} \Sigma_{j=1}^{J} n_{ij}$. Now the unbiased estimator of q_{ijk} is Y_{ijk}/n_{ij}, so the mean midrank for the ith sample is

$$\overline{R}_{ij} = \sum_{k=1}^{K} \bar{r}_k \hat{q}_{ijk}. \quad (4.21)$$

We call each $\bar{r}_k$ the rank score, and the null hypothesis of interest is

$$H_0: \quad E(\overline{R}_{11}) = E(\overline{R}_{12}) = \cdots = E(\overline{R}_{IJ}). \quad (4.22)$$

If H_0 is false, we can form various models to characterize the variation in the R_{ij}, just as we did when analyzing the q_{ij} under models of homogeneity.

Selvin (1977) illustrated the relationship between the normed mid-ranks and ridits using data on socioeconomic status (SES), age, and smoking (Table 4.3). The factor-response structure is not evident from the discussion of the study, but for purposes of illustration, assume that SES is the response and age and smoking status are factors. The global hypothesis is no difference in the mean midrank for each of the six groups. More specifically, we are interested in whether the median SES of the three smoking groups is the same after adjusting for age effects.

The first step in the analysis is to compute the rank scores, $\bar{r}_k$. Using the row labeled total we apply (4.20):

$$\bar{r}_1 = \frac{78 + 1}{2(740)} = 0.053$$

$$\bar{r}_2 = \frac{78 + (328 + 1)/2}{740} = 0.328$$

$$\vdots$$

$$\bar{r}_5 = \frac{78 + 328 + 141 + 122 + (71 + 1)/2}{740} = 0.953.$$

These scores are applied in each of the six groups to obtain the mean midranks using (4.21). For example,

$$\overline{R}_{11} = 0.053(0.058) + 0.328(0.500) + \cdots + 0.953(0.084) = 0.497.$$

Table 4.3 Socioeconomic Status by Age and Smoking Status

				Socioeconomic Status					
Age (i)	Smoking Status (j)	(i, 1, k)		Low (i, j, 1)	2 (i, j, 2)	3 (i, j, 3)	4 (i, j, 4)	High (i, j, 5)	Total (i, j, +)
<55	Never	(1, 1, k)	n	9	77	38	17	13	154
		%	5.8	50.0	24.7	11.0	8.4	100.0	
	Former	(1, 2, k)	n	1	10	3	5	6	25
		%	4.0	40.0	12.0	20.0	24.0	100.0	
	Present	(1, 3, k)	n	14	84	44	26	17	185
		%	7.6	45.4	23.8	14.1	9.2	100.0	
55+	Never	(2, 1, k)	n	37	129	41	51	14	272
		%	13.6	47.4	15.1	18.8	5.1	100.0	
	Former	(2, 2, k)	n	2	6	3	6	7	24
		%	8.3	25.0	12.5	25.0	29.2	100.0	
	Present	(2, 3, k)	n	15	22	12	17	14	80
		%	18.8	27.5	15.0	21.2	17.5	100.0	
Total[a]		(+, +, k)	n	78	328	141	122	71	740
		%	10.5	44.3	19.0	16.5	9.6	100.0	

[a]Totals differ because of rounding.
Source: Selvin (1977).

The remaining mean midranks are shown in the second column of Table 4.4. Using the simple mean scoring procedures of Section 3.3.1, we have also generated the mean SES, $\bar{y}_{ij}$, for each group, shown in the first column of Table 4.4.

The variances for the $\overline{R}_{ij}$ are estimated using the Neyman procedure discussed in the Appendix (4A.1). In this application the response function has the simple form $\mathbf{f}(\hat{\mathbf{q}}_{ij}) = \mathbf{a}'\hat{\mathbf{q}}_{ij}$, where $\mathbf{a}' = (\bar{r}_1, \bar{r}_2, \ldots, \bar{r}_5) = (0.053, 0.328, 0.645, 0.822, 0.953)$ and $\hat{\mathbf{q}}_{ij}' = (\hat{q}_{ij1}, \hat{q}_{ij2}, \ldots, \hat{q}_{ij5})$.

The next step in the analysis is to propose a model for the scores. Since only two factors are involved, a main-effect model is a reasonable choice:

$$E(f(\hat{p}_{ij})) = M + X_1 A + X_2 S_1 + X_3 S_2,$$

where

$$X_1 = -1 \text{ when } i = 1\ (<55), \quad 1 \text{ when } i = 2\ (55+)$$

$$X_2 = 1 \text{ when } j = 1 \text{ (never)}, \quad 0 \text{ when } j = 2 \text{ (former)}, \quad -1 \text{ when } j = 3 \text{ (present)}$$

$$X_3 = 0 \text{ when } j = 1 \text{ (never)}, \quad 1 \text{ when } j = 2 \text{ (former)}, \quad -1 \text{ when } j = 3 \text{ (present)}.$$

Table 4.4 Mean Scores, Mean Midranks, and Ridits[a]

		Score		
Age i	Smoking Status j	$\overline{Y}_{ij}$	Midranks R_{ij}	Ridits S_{ij}
<55	Never	2.66	0.497	0.536
	Former	3.20	0.604	0.639
	Present	2.72	0.509	0.547
55+	Never	2.54	0.463	0.500
	Former	3.42	0.650	0.682
	Present	2.91	0.538	0.570

[a]Ridits relative to 55+, never-smoked sample.

Here the dummy variable for age has been constructed exactly as before, but since smoking has three levels, two dummy variables (X_2, X_3) are required. As before, this parameterization is quite arbitrary and affects only the interpretation of the model parameters.

The test statistics and estimates are shown in Table 4.5. First, it is clear from the residual Q-statistics (df = 2) that the data fit the model whether mean scores or mean midranks are used. Next, for both response functions it is clear that age has no significant association with SES, but smoking is significant. The parameter estimates allow us to interpret this relationship. Since never and former smokers are contrasted with present smokers and the parameter signs are negative and positive, respectively, it follows that having never smoked is associated with generally lower SES level than present smokers, but former

Table 4.5 Test Statistics for Mean Scores, Mean Midranks, and Ridits

a. Analysis of Variation

		Mean Score		Midrank		Ridit	
Source	df	Q	p-value	Q	p-value	Q	p-value
Model	3						
Age	1	0.05	0.821	0.35	0.556	0.55	0.457
Smoking	2	14.25	0.001	12.10	0.002	11.56	0.003
Residual	2	2.73	0.255	2.23	0.327	2.06	0.356
Total	5						

b. Estimated Parameters

		Mean Score		Midrank		Ridit	
Interpretation	Label	Estimate	SE	Estimate	SE	Estimate	SE
Intercept	M	2.88	0.07	0.54	0.02	0.57	0.02
55+ vs. <55	A	−0.01	0.04	−0.01	0.01	−0.01	0.01
Never vs. present	S_1	−0.29	0.08	−0.06	0.02	−0.06	0.01
Former vs. present	S_2	0.42	0.13	0.09	0.03	0.08	0.03

smokers are associated with a higher SES level. These conclusions hold for both response functions.

Mean scores are generally associated with an underlying continuum and ranks are not readily interpretable, especially when one wants to compare several studies. For this reason, Bross (1958) proposed using the cumulative distribution of some identified populations as a scoring scheme. As noted by Selvin (1977), if the marginal distribution of the response is used as the identified population, ridit scores are equivalent to the mean midrank scores.

The ridit scores are defined as follows. Let the proportion of the identified population at the kth response level be denoted s_k; the ridit score is defined to be

$$\bar{s}_k = \sum_{k'=1}^{k-1} \frac{s_{k'} + s_k}{2}. \tag{4.24}$$

The mean ridit score for group (i,j) is then

$$\bar{S}_{ij} = \sum_{k=1}^{K} \bar{s}_k \hat{q}_{ijk} \tag{4.25}$$

Now suppose that $\bar{s}_k = Y_{++k}/n$, so

$$\bar{s}_k = \frac{\Sigma_{k'=1}^{k-1} Y_{++k'}}{n} + \frac{Y_{++k}}{2n} = \frac{\Sigma_{k'=1}^{k-1} Y_{++k'}}{n} + \frac{Y_{++1} + 1}{2n} - \frac{1}{2n}$$

$$= \frac{\Sigma_{k'=1}^{k-1} Y_{++k'} + (Y_{++k} + 1)/2}{n} - \frac{1}{2n} = \bar{r}_k - \frac{1}{2n}.$$

Hence $\bar{s}_k$ is the mean midrank reduced by $1/2n$. Of course, in Bross's original formulation s_k was selected to correspond to some appropriate standard.

To illustrate the procedure, suppose that persons 55+ who have never smoked are selected as the standard. Apply (4.24), the scores are $\bar{s}_1 = 0.136/2 = 0.068$, $\bar{s}_2 = 0.136 + (0.474/2) = 0.373, \ldots, \bar{s}_5 = 0.974$. These scores applied to each sample yield the mean ridits shown in the final column of Table 4.4. Notice that in the identified population the mean ridit is 0.5, which it must be. Again applying

(4A.5), where $a_k = s_k$, the variances of the ridits are immediately obtained.

The analysis proceeds by estimating the main-effect model (4.23) and computing the appropriate test statistics. The results are shown in the final two columns of Table 4.5. The conclusions are the same as those made using means and mean midrank scores, as one would expect.

4.3 LOG-LINEAR MODEL IN CONTINGENCY TABLES

The models for homogeneity in Section 4.1 and 4.2 have three components:

1. Response variables
2. Appropriate functions of the response variables
3. A set of dummy variables to characterize the underlying study design

This detailed set of specifications requires that the analyst understand not only the conceptual structure of the study hypotheses, but also the method of data collection. The design that led to the plum root stalk data set (Table 4.1) is reasonably obvious, but the smoking–age–SES data set (Table 4.3) required a number of assumptions in order to validly treat socioeconomic status as an ordered response variable. For many data sets in the health, medical, and social sciences, the factor-response structure is either absent or difficult to ascertain. This is particularly true for cross-sectional studies.

A further complication occurs when case-control designs are employed. If we follow the study design, the presence of the disease is the factor ("case-ness") and all measured variables are responses, so the factor-response structure is reversed. Implicitly, a researcher wishes to estimate the conditional probabilities of being either a case or control, but all that can be estimated directly is the conditional probability of the response given case or control status. As demonstrated in Section 2.1.3, when the probability of observing a case in general is small ("rare disease assumption"), the odds ratio can be utilized as an approximation of the relative risk.* Moreover, if the choice of depen-

*This assumption is sufficient but not necessary (e.g., Kleinbaum et al. 1982).

dent variable is not obvious, the invariance of the odds ratio to row and column reversals means that it can be utilized for studies of independence.

For these reasons, and others, the odds ratio has been utilized for most studies with a multiple-response structure. The generalization of the odds ratio to multidimensional tables is usually based on the "log-linear model" (Bishop et al. 1975; Fienberg 1980; Goodman 1978; Haberman 1975, 1978, 1979). These models are analysis-of-variance (ANOVA) models applied to the natural logarithms of the multinomial probabilities or expected cell counts. In addition, a particular parameterization is usually employed. Nearly all researchers on the log-linear model have relied on maximum likelihood estimators (m.l.e.'s). It is possible, but somewhat cumbersome, to develop a parallel set of weighted least squares estimates. The latter approach is discussed by Forthofer and Lehnen (1981), Grizzle and Williams (1972), and Imrey et al. (1981, 1982). We will utilize the m.l.e. approach because of the widespread availability of appropriate and convenient software (e.g. Dixon 1981).

The log-linear model has a special appeal because of its apparent similarity to ANOVA models. However, interpretations of the model are sometimes elusive. We begin the discussion by examining it in a 2×2 table and then extending the model to an $I \times J$ table (Section 4.3.2). Although the model is somewhat cumbersome in this context, the interpretation will generalize to three-way tables, which are considered in Section 4.4.

4.3.1 2 × 2 Tables

The log-linear model for a 2×2 table can be illustrated by treating the CHD-tension data in Table 4.6 as if they were collected cross-sectionally. The corresponding table of underlying parameters for the distribution is Table 3.4. Again, we make the assumption of random sampling, which leads to the cell counts Y_{ij}, following a multinomial distribution with parameters n and p_{ij}, where $P(T = i \text{ and } C = j) = p_{ij} > 0$, $i = 1, 2$, $j = 1, 2$, and $T = 1, 2$ for tension (yes or no), and $C = 1, 2$ for CHD (yes or no). Let the natural logarithm of the expected cell count be given by

$$l_{ij} = \ln [E(Y_{ij})] = \ln (m_{ij}) = \ln (n) + \ln (p_{ij}). \qquad (4.26)$$

Table 4.6 Major CHD Event by Report of Job Tension Observed and Expected Counts for 2013 Western Electric Workers

Do You Work Under Tension?	Major CHD Event Yes	Major CHD Event No	Total
Yes	97	307	404
	m_{11}	m_{12}	m_{1+}
No	200	1409	1609
	m_{21}	m_{22}	m_{2+}
Total	297	1716	2013
	m_{+1}	m_{+2}	$m_{++} = n$

Source: Ostfeld (1980).

Note that since $m_{ij} = np_{ij}$, models can be written in terms of either expected counts or underlying probabilities. We use the former since the corresponding estimators are in terms of the observed counts, y_{ij}.

The saturated log-linear model is entirely analogous to what was used for the two-factor, one-response model (4.5):

$$l_{ij} = \ln (m_{ij}) = M + T_i + C_j + (TC)_{ij} \tag{4.27}$$

where $\Sigma_i T_i = \Sigma_j C_j = \Sigma_i (TC)_{ij} = \Sigma_j (TC)_{ij} = 0$ are the usual identifiability conditions. Following the algebra of Section 4.1, we obtain $M = \Sigma_i \Sigma_j l_{ij}/4 = l_{++}/4$. In the example, this leads to the m.l.e.:

$$\hat{M} = \frac{\ln (97) + \ln (307) + \ln (200) + \ln (1409)}{4} = 5.713.$$

Similarly, the row and column effects are

$$T_i = \frac{\Sigma_j l_{ij}}{2} - \frac{l_{++}}{4} = \frac{l_{i+}}{2} - \frac{l_{++}}{4}, \tag{4.28}$$

$$C_j = \frac{\Sigma_i l_{ij}}{2} - \frac{l_{++}}{4} = \frac{l_{+j}}{2} - \frac{l_{++}}{4}. \tag{4.29}$$

Substituting the expected counts for l_{ij} we obtain for $i = 1$

$$T_i = \frac{\ln(m_{11}) + \ln(m_{12})}{2}$$

$$- \frac{\ln(m_{11}) + \ln(m_{12}) + \ln(m_{21}) + \ln(m_{22})}{4}$$

$$= \frac{\ln(m_{11}) - \ln(m_{21}) + \ln(m_{12}) - \ln(m_{22})}{4}$$

$$= \frac{\ln(m_{11}/m_{21})(m_{12}/m_{22})}{4}.$$

Similarly, the coronary heart disease main effect at $j = 1$ is

$$C_1 = \frac{\ln(m_{11}/m_{12})(m_{21}/m_{22})}{4}.$$

Again, the m.l.e. just uses the y_{ij} in place of the m_{ij}. In our example

$$\hat{T}_1 = \frac{\ln(97/200)(307/1409)}{4} = \frac{\ln(0.1057)}{4} = -0.562$$

$$\hat{C}_1 = \frac{\ln(97/307)(200/1409)}{4} = \frac{\ln(0.0448)}{4} = -0.776.$$

Because of the identifiability conditions (4.27), $T_2 = -T_1$ and $C_2 = -C_1$, so only three parameters have been defined so far: an overall mean and two main effects.

The fourth and last parameter is the deviation of the first three parameters from the individual cell values:

$$(TC)_{ij} = l_{ij} - T_i - C_j - M = l_{ij} - \frac{l_{i+} + l_{+j}}{2} + \frac{l_{++}}{4}. \tag{4.30}$$

For example, $\hat{TC}_{11} = \ln(97) - (-0.562) - (-0.776) - (5.713) = 0.200$. Now recall that the log-odds ratio for this table is $\ln(\hat{\text{OR}}_{11}) = \ln[97(1409)/200(307)] = 0.800 = 4(\hat{TC}_{11})$. This is an illustration of the general result that the "interaction parameter" of the log-linear model is a function of the corresponding odds ratio.

Once the interaction term is recognized as being proportional to the logarithm of the odds ratio, we have a key to understanding and interpreting all the parameters of the log-linear model. In Section 3.5 we observed that two-response tables can be converted into one-factor, one-response tables by conditioning on the appropriate margin. For example, among workers who suffer a major CHD event, the conditional proportion who reported working under tension is $q_{11} = p_{11}/p_{+1} = m_{11}/m_{+1}$. This is estimated as $\hat{q}_{11} = 97/297 = 0.327$. Similarly, the conditional *odds* of reporting working under tension given the worker is among those who will subsequently suffer a CHD event is $\mathrm{OT}_1 = q_{11}/(1 - q_{11}) = m_{11}/m_{21}$. The corresponding estimate is $\hat{\mathrm{OT}}_1 = 97/200 = 0.485$. The same reasoning leads to observing that the odds of working under tension among those who do not subsequently suffer a CHD event is $\mathrm{OT}_2 = m_{12}/m_{22}$. These two odds are the components of the log-linear model main-effect parameter:

$$T_1 = \frac{\ln\,[(m_{11}/m_{21})(m_{12}/m_{22})]}{4} = \frac{\ln\,[\mathrm{OT}_1(\mathrm{OT}_2)]}{4}.$$

That is, the parameter is written as the log odds of reporting working under tension multiplied across the levels of C. Similarly, the log odds of coronary heart disease multiplied across the levels of tension yields the other main effect:

$$C_1 = \frac{\ln\,[(m_{11}/m_{12})(m_{21}/m_{22})]}{4} = \frac{\ln\,[\mathrm{OC}_1(\mathrm{OC}_2)]}{4}.$$

The relationship between the model parameters and the table odds ratios can be illustrated by selecting one level of each variable as a reference value. For example, suppose that $i = j = 1$ are the reference values. The odds that $i = 1$ and $j = 1$ are given by $\mathrm{OT}_j = m_{1j}/m_{2j}$ and $\mathrm{OC}_i = m_{i1}/m_{i2}$. Here OT_j is the odds that $T = 1$ at level j of C. Similarly, OC_i is the odds that $C = 1$ at level i of tension. The odds ratio for $i = 1, j = 1$ is then

$$\mathrm{OR}_{11} = \frac{\mathrm{OT}_1}{\mathrm{OT}_2} = \frac{\mathrm{OC}_1}{\mathrm{OC}_2}. \tag{4.31}$$

The log-linear model for the cell ($i = 1, j = 1$) is

$$l_{11} = \ln(m_{11}) = M + T_1 + C_1 + TC_{11}$$
$$= M + \frac{\ln[\mathrm{OT}_1(\mathrm{OT}_2)]}{4} + \frac{\ln[\mathrm{OC}_1(\mathrm{OC}_2)]}{4} + \frac{\ln(\mathrm{OR}_{11})}{4}.$$

Since $T_2 = -T_1$, $C_2 = -C_2$, and $TC_{11} = -TC_{12} = -TC_{21} = TC_{22}$, all the remaining l_{ij} are also functions of the corresponding odds and odds ratios:

$$l_{12} = \frac{4M + \ln[\mathrm{OT}_1(\mathrm{OT}_2)] - \ln[\mathrm{OC}_1(\mathrm{OC}_2)] - \ln(\mathrm{OR}_{11})}{4},$$

$$l_{21} = \frac{4M - \ln[\mathrm{OT}_1(\mathrm{OT}_2)] + \ln[\mathrm{OC}_1(\mathrm{OC}_2)] - \ln(\mathrm{OR}_{11})}{4},$$

$$l_{22} = \frac{4M - \ln[\mathrm{OT}_1(\mathrm{OT}_2)] - \ln[\mathrm{OC}_1(\mathrm{OC}_2)] + \ln(\mathrm{OR}_{11})}{4}.$$

Now suppose that we wished to test the hypothesis of independence:

$$H_0: \quad m_{ij} = \frac{m_{i+}m_{+j}}{n}. \tag{4.32}$$

We already know that this implies that $\mathrm{OR}_{11} = 1$, hence $TC_{ij} = 0$. As one might anticipate, it is also true that $TC_{ij} = 0$ implies that $\mathrm{OR}_{11} = 1$ and hence H_0. This can be seen by examining the log-linear model main-effect parameters when the hypothesis of independence (4.32) is true:

$$4T_i = [2\ln(m_{i1}) + 2\ln(m_{i2})$$
$$- \ln(m_{11}m_{12}m_{21}m_{22})]$$
$$= 2[\ln(m_{i+}) + \ln(m_{+1}) - \ln(n)]$$
$$+ 2[\ln(m_{i+}) + \ln(m_{+2}) - \ln(n)]$$
$$- [\ln(m_{1+}) + \ln(m_{+1}) - \ln(n)]$$

$$- [\ln(m_{1+}) + \ln(m_{+2}) - \ln(n)]$$

$$- [\ln(m_{2+}) + \ln(m_{+1}) - \ln(n)]$$

$$- [\ln(m_{1+}) + \ln(m_{+2}) - \ln(n)]$$

$$4T_i = 4\ln(m_{i+}) - 2\ln(m_{1+}) - 2\ln(m_{2+})$$

$$\text{so } T_i = \ln(m_{i+}) - \frac{\ln(m_{1+}m_{1+})}{2}.$$

Similarly, $C_j = \ln(m_{+}j) - \ln(m_{+1}m_{+2})/2$ and $M = \ln(m_{1+}m_{2+}m_{+1}m_{+2})/2 - \ln(n)$. Combining these, the log-linear model under independence is

$$l_{ij} = \ln(m_{ij}) = M + T_i + C_j$$

$$= \frac{\ln(m_{1+}m_{2+}m_{+1}m_{+2})}{2} + \frac{\ln(m_{i+}) - \ln(m_{1+}m_{2+})}{2}$$

$$+ \frac{\ln(m_{+}j) - \ln(m_{+1}m_{+2})}{2}$$

$$= \ln(m_{i+}m_{+}j) - \ln(n) = \ln\left(\frac{m_{i+}m_{+j}}{n}\right).$$

Hence the log-linear model when $TC_{ij} = 0$ is just another way of stating (4.32). But more important, when $TC_{ij} = 0$, the m.l.e. for the log-linear model can be obtained from the margins of the table. This is a simple example of what is called *collapsibility*. We say that T_i is collapsible with respect to C_j (and conversely), when T_i can be validly estimated without regard to C_j (and conversely).

4.3.2 $I \times J$ Tables

The log-linear model readily extends to rectangular tables, that is, tables where $I \geq 2$ and $J \geq 2$. As a simple example, consider the 4×2 table on the sex distribution for various villages of the Yanomamö Indians of the Upper Amazon Basin (Table 4.7). Computing the usual chi-square test of independence indicates that there is a significant relationship between the lineage and the sex distribution. Specifically, there is an excess of males among the Hor and a deficit among the Shamatar. The sex ratio or

Table 4.7 Sex Distribution for Various Lineages of Several Yanomamö Villages[a]

		Sex				
		Male		Female		
Lineage	Observed Expected	Count $(i, 1)$	Odds $(i, 1)$	Count $(i, 2)$	Odds $(i, 2)$	Total Count $(i, +)$
Sha	$(1, j)$	70	1.19	59	0.84	129
		67.93	1.11	61.07	0.90	
Hor	$(2, j)$	82	1.71	48	0.58	130
		68.45	1.11	61.55	0.90	
Shamatar	$(3, j)$	11	0.35	31	2.82	42
		22.12	1.11	19.88	0.90	
Other	$(4, j)$	55	0.95	58	1.06	113
		59.50	1.11	53.50	0.90	
Total	$(+, j)$	218	1.11	196	0.90	414

$X^2 = 18.32$ $G^2 = 18.77$ df = 3

[a]Counts, estimated expected values, odds, and odds under independence.
Source: Chagnon (1977, Table 3–1, p. 67).

ratio of males to females is a common measure of the sex distribution. This is shown in the "odds" column for males in Table 4.7. Overall, the value is 1.11, but the sex ratio for the Hor and Shamatar deviate substantially from this value. We can again define a log-linear model to characterize the variation in this table. Let Y_{ij} correspond to the multinomial random variable in cell (i,j), $i = 1 \ldots, 4$, $j = 1, 2$. As before, let

$$l_{ij} = \ln [E(Y_{ij})] = \ln (m_{ij}) = \ln (p_{ij}) + \ln (n), \tag{4.33}$$

where $p_{ij} = P(i\text{th lineage and } j\text{th sex}) > 0$. Again by analogy to analysis of variance a model for these log expected counts is $l_{ij} = M + L_i + S_j + (LS)_{ij}$, where $\Sigma_i L_i = \Sigma_j S_j = \Sigma_i (LS)_{ij} = \Sigma_j (LS)_{ij} = 0$.

The overall mean is

$$M = \frac{\sum_{i=1}^{4} \sum_{j=1}^{2} l_{ij}}{8} = \frac{l_{++}}{8}. \tag{4.34}$$

This is estimated by

$$\hat{M} = \frac{\ln [70(59)(82)(48)(11)(31)(55)(58)]}{8} = 3.813.$$

Similarly, the lineage (L_i) and sex (S_j) effects are deviations of the l_{ij} averages from this mean:

$$\begin{aligned} L_i &= \sum_{j=1}^{2} \frac{l_{ij}}{2} - \frac{l_{++}}{8} = \frac{l_{i+}}{2} - \frac{l_{++}}{8} \\ S_j &= \sum_{i=1}^{4} \frac{l_{ij}}{4} - \frac{l_{++}}{8} = \frac{l_{+j}}{4} - \frac{l_{++}}{8} \end{aligned} \tag{4.35}$$

For example, the deviation of males from the average is

$$\begin{aligned} S_1 &= \frac{l_{11} + l_{21} + l_{31} + l_{41}}{4} \\ &\quad - \frac{l_{11} + l_{21} + l_{31} + l_{41} + l_{12} + l_{22} + l_{32} + l_{42}}{8} \\ &= \frac{(l_{11} - l_{12}) + (l_{21} - l_{22}) + (l_{31} - l_{32}) + (l_{41} - l_{42})}{8}. \end{aligned}$$

The corresponding estimate is one-eighth the sum of the log sex ratios:

$$\hat{S}_1 = \frac{\ln (70/59) + \ln (82/48) + \ln (11/31) + \ln (55/58)}{8} = -0.048.$$

Since $S_2 = -S_1$, it follows that $\hat{S}_2 = 0.048$.

This "averaging" of log odds is the same as in Section 4.3.1 for the 2 × 2 table. The term "average" is in quotes to emphasize that it is not exactly an average but a sum divided by an appropriate constant.

The estimators differ from what we have done so far when more than two levels must be considered. As an example, consider the village main effect for the Shamatar:

$$L_3 = \frac{l_{31} + l_{32}}{2} - \frac{l_{11} + l_{12} + l_{21} + l_{22} + l_{31} + l_{32} + l_{41} + l_{42}}{8}$$

$$= \frac{3l_{31} + 3l_{32} - l_{11} - l_{12} - l_{21} - l_{22} - l_{41} - l_{42}}{8} .$$

We can gain some insight into the interpretation of such a complicated function by rewriting it in terms of the expected counts:

$$\begin{aligned} L_3 &= [3 \ln (m_{31}) + 3 \ln (m_{32}) - \ln (m_{11}) - \ln (m_{12}) - \\ &\quad \ln (m_{21}) - \ln (m_{22}) - \ln (m_{41}) - \ln (m_{42})]/8 \\ &= [\ln (m_{31}/m_{11}) + \ln (m_{31}/m_{21}) + \ln (m_{31}/m_{41}) + \\ &\quad \ln (m_{32}/m_{12}) + \ln (m_{32}/m_{22}) + \ln (m_{32}/m_{42})]/8 \end{aligned}$$

Now consider the components of L_3. These are of the form m_3j/m_{ij}, which while not being an odds is certainly similar to one. Thus L_3 can be thought of as a type of average log odds of being Shamatar as opposed to all other lineages. The corresponding estimate is

$$\begin{aligned} \hat{L}_3 &= [3 \ln (11) + 3 \ln (31) - \ln (70) - \ln (59) - \\ &\quad \ln (82) - \ln (48) - \ln (55) - \ln (58)]/8 = -0.897. \end{aligned}$$

A similar computation yields $\hat{L}_1 = 0.350$, $\hat{L}_2 = 0.326$, and $\hat{L}_4 = 0.221$. Notice that $\Sigma_{i=1}^{4} \hat{L}_i = 0$.

Finally, the interaction terms are defined to be the deviation of the model parameters from the log expected cell counts:

$$(LS)_{ij} = l_{ij} - (m + L_i + S_j) = l_{ij} - \left(\frac{l_{i+}}{2} + \frac{l_{+j}}{4}\right) + \frac{l_{++}}{8}. \quad (4.36)$$

For example, the interaction term for a female Shamatar ($i = 3, j = 2$) is

$$\begin{aligned}(LS)_{32} &= l_{32} - \left(\frac{l_{3+}}{2} + \frac{l_{+2}}{4}\right) + \frac{l_{++}}{8} \\ &= \ln(m_{32}) - \frac{\ln(m_{31}) + \ln(m_{32})}{2} \\ &\quad - \frac{\ln(m_{12}) + \ln(m_{22}) + \cdots + \ln(m_{42})}{4} \\ &\quad + \frac{\ln(m_{11}) + \cdots + \ln(m_{42})}{8} \\ &= [8\ln(m_{32}) - 4[\ln(m_{31}) + \ln(m_{32})] - 2[\ln(m_{12}) + \ln(m_{22}) + \\ &\quad \ln(m_{32}) + \ln(m_{42})] + \ln(m_{11}) + \ln(m_{12}) + \cdots + \ln(m_{42})]/8 \\ &= [3\ln(m_{32}) - \ln(m_{12}) - \ln(m_{22}) - \ln(m_{42}) - \\ &\quad 3\ln(m_{32}) + \ln(m_{11}) + \ln(m_{21}) + \ln(m_{42})]/8\end{aligned}$$

$$(LS)_{32} = \frac{\ln(m_{32}m_{11}/m_{31}m_{12}) + \ln(m_{32}m_{21}/m_{31}m_{22}) + \ln(m_{32}m_{32}/m_{31}m_{41})}{8}.$$

That is, $(LS)_{32}$ is the "average" of log odds ratios using female Shamatar ($i = 3$, $j = 2$) as the reference cell.

The estimators are obtained by replacing the m_{ij} by the observed Y_{ij}. For example,

$$\begin{aligned}(\hat{LS})_{32} &= \frac{3\ln(31) + \ln[70(82)(55)] - 3\ln(11) - \ln[59(48)(58)]}{8} \\ &= 0.470.\end{aligned}$$

We get the same result from (4.36): $(LS)_{32} = \ln(31) - (3.813 + 0.048 - 0.897) = 0.470$. Similar computations yield $(LS)_{12}$

$= -0.133$, $(LS)_{22} = -0.316$, $(LS)_{42} = -0.021$. Since $\Sigma_j (LS)_{ij} = 0$, the male values are the same in absolute value but have the opposite sign. Finally, notice that $(LS)_{22}$ is relatively small and $(LS)_{32}$ relatively large; this corresponds to the earlier observation concerning a deficit of females (excess of males) among the Hor and an excess of females (deficit of males) among the Shamatar. The log-linear model allows us to characterize explicitly the variation among the table cells.

The next question concerns the hypothesis of independence:

$$H_0: \quad m_{ij} = \frac{m_{i+}m_{+j}}{n}, \qquad i = 1, \ldots, I, \quad j = 1, \ldots, J. \tag{4.37}$$

As with the 2×2 table a little algebra shows that lineage and sex are independent if and only if $(LS)_{ij} = 0$ for all i, j. Therefore, the hypothesis can be stated as

$$H_0: \quad (LS)_{ij} = 0. \tag{4.38}$$

Since $\Sigma_i (LS)_{ij} = \Sigma_j (LS)_{ij} = 0$, there are only three parameters associated with this hypothesis. This corresponds to the three degrees of freedom for the chi-square test statistics.

The test of independence or no "interaction" between lineage and sex is conducted by estimating the expected cell frequencies given the model

$$l_{ij} = M + L_i + S_j. \tag{4.39}$$

As was the case for the 2×2 table it can be shown that the estimates of the model parameters now depend only on the marginal frequencies:

$$M = \frac{2\,\Sigma_i \ln(m_{i+}) + 4\,\Sigma_j \ln(m_{+j}) - 8 \ln(n)}{8} \tag{4.40}$$

$$L_i = \ln(m_{i+}) - \frac{\Sigma_i \ln(m_{i+})}{4} \qquad S_j = \ln(m_{+j}) - \frac{\Sigma_j \ln(m_{+j})}{2}.$$

The general log-linear model for the contingency table for the responses A and B, with I and J levels, respectively, is defined by

$$l_{ij} = M + A_i + B_j + (AB)_{ij} \tag{4.41}$$

where $l_{ij} = \ln(m_{ij}) = \ln(p_{ij}) + \ln(n)$, $M = \Sigma_{i=1}^{I}\Sigma_{j=1}^{J} l_{ij}/IJ = l_{++}/IJ$, $A_i = \Sigma_{j=1}^{J} l_{ij}/J - l_{++}/IJ$, $B_j = \Sigma_{i=1}^{I} l_{ij}/I - l_{++}/IJ$, $(AB)_{ij} = l_{ij} - (m + A_i + B_j)$. Here $(AB)_{ij}$ can be interpreted as the average log odds ratio for cell (i, j) compared to all other cells. The remaining terms correspond to the average log odds for the indicated category. The hypothesis of independence corresponds to the hypothesis of no "interaction" $[(AB)_{ij} = 0]$. If the variables A and B are independent, then A_i is collapsible with respect to B, and conversely.

This notation is somewhat simpler than that found in most texts on log-linear models. The translation to what are sometimes called *u-terms* is given by

$$\theta = M \qquad u_{1(i)} = A_i \qquad u_{2(j)} = B_j \qquad u_{12(ij)} = AB_{ij}. \tag{4.42}$$

4.4 THE LOG-LINEAR MODEL FOR THREE-VARIABLE TABLES

The log-linear model has the advantage of generalizing to tables defined by more than two variables without requiring that any variable be specified as the response. This is particularly useful when all the variables are observed simultaneously or when the underlying structure of the data is not obvious to the investigator. Bishop (1969) first reported the data on infant survival shown in Table 4.8. Here the survival of infants whose mothers received either less than 1 month, or 1 or more months of care in one of two clinics is tabulated for 715 infants. Because the survival outcome must occur after the prenatal care and clinic categorization, it could be argued that this is a two-factor, one-response problem and a model of homogeneity is appropriate. However, as noted by Bishop (1969, p. 388): "their [the data] original source is not known." So a model based on three responses is more appropriate for analyzing the data. This corresponds to modeling

Table 4.8 Infant Survival According to Duration of Prenatal Care Received by the Mothers and Clinic

			Infant Survival, S				
	Months of Prenatal		Died $(i, j, 1)$		Alive $(i, j, 2)$		
Clinic (C)	Care (D)	Count	Odds	Count	Odds	Count	Total $(i, j, +)$
A	<1	$(1, 1, k)$	3	0.017	176	58.667	179
	1+	$(1, 2, k)$	4	0.014	293	73.250	297
B	<1	$(2, 1, k)$	17	0.086	197	11.588	214
	1+	$(2, 2, k)$	2	0.087	23	11.500	25
Total	<1	$(+, 1, k)$	20	0.054	373	18.650	393
	1+	$(+, 2, k)$	6	0.019	316	52.667	322
Total		$(+, +, k)$	26	0.038	689	26.500	715

Source: Bishop et al. (1975).

patterns of independence among the three variables. As noted in Section 4.3, models of independence are equivalent to constructing models based on odds ratios.

The analysis begins, as in Section 4.3, by considering the tabulation of duration of prenatal care (D) and survival (S). Using either the Pearson chi-square ($X^2 = 5.26$) or the likelihood ratio chi-square ($G^2 = 5.61$), we conclude there is evidence of a statistical association. Let $j = 1, 2$ index amount of care and $k = 1, 2$ index survival. A log-linear model for this 2×2 table is given by

$$\ln (m_{+jk}) = M_{.} + D_{.j} + S_{.k} + (DS)_{.jk} \qquad \text{where } m_{+jk} = \sum_i m_{ijk} \quad (4.43)$$

and the parameters are defined as in (4.41). The "." subscript indicates that the parameters are based on the table collapsed or summed over the omitted index ($i = 1, 2$ for clinic). The estimates for this model are obtained as before (4.27–4.30):

$$\hat{M}_{.} = \frac{\ln [(20)(373)(6)(316)]}{4} = 4.12$$

$$\hat{D}_{.1} = \frac{\ln [(20)(373)/(6)(316)]}{4} = 0.34$$

$$\hat{S}_{.1} = \frac{\ln [(20)(6)/(373)(316)]}{4} = -1.72$$

$$(\hat{DS})_{.11} = \frac{\ln [(20)(316)/(6)(373)]}{4} = 0.26.$$

The interaction, $(\hat{DS})_{.11}$, is positive, so on purely statistical grounds we infer that survival is associated with more care in the collapsed 2 × 2 table.

Inspection of Table 4.8 stratified by clinic suggests that this conclusion does not hold within each clinic. Notice that in clinic 1 the survival odds are 58.667 and 73.250, and in clinic 2 the odds are 11.588 and 11.500. Clearly, there is little or no effect in clinic 2 and not a marked effect in clinic 1. Moreover, there is a large difference in the survival odds between the two clinics. This suggests that we construct log-linear models for each of the clinics separately. Let $i = 1, 2$ index the two clinics. Then we have

$$\begin{aligned} \ln (m_{1jk}) &= M_1 + D_{1j} + S_{1k} + (DS)_{1jk} \\ \ln (m_{2jk}) &= M_2 + D_{2j} + S_{2k} + (DS)_{2jk}. \end{aligned} \tag{4.44}$$

These can be combined to produce one model for both tables.

First consider the sum and difference of (4.44):

$$\begin{aligned} \ln (m_{1jk}) + \ln (m_{2jk}) &= (M_1 + M_2) + (D_{1j} + D_{2j}) + (S_{1k} + S_{2k}) \\ &\quad + [(DS)_{1jk} + (DS)_{2jk}] \\ \ln (m_{1jk}) - \ln (m_{2jk}) &= (M_1 - M_2) + (D_{1j} - D_{2j}) + (S_{1k} - S_{2k}) \\ &\quad + [(DS)_{1jk} - (DS)_{2jk}]. \end{aligned} \tag{4.45}$$

Next we average the two equations in (4.45):

$$\ln (m_{1jk}) = \frac{(M_1 + M_2) + (M_1 - M_2)}{2} + \frac{(D_{1j} + D_{2j}) + (D_{1j} - D_{2j}) + (S_{1j} + S_{2j}) + (S_{1j} - S_{2j})}{2} + \frac{(DS)_{1jk} + (DS)_{2jk} + (DS)_{1jk} - (DS)_{2jk}}{2}. \tag{4.46}$$

From this combined equation we can generate all the log-linear model parameters for a 2 × 2 × 2 table.

Letting $\ln (m_{ijk}) = l_{ijk}$, we rewrite each component of the right-hand side of (4.46) to obtain model parameters for the three-way table:

$$\frac{(M_1 + M_2)}{2} = \frac{l_{1++}/4 + l_{2++}/4}{2} = \frac{l_{+++}}{8} = M$$

$$\frac{D_{1j} + D_{2j}}{2} = \frac{(l_{1j+}/2 - l_{1++}/4) + (l_{2j+}/2 - l_{2++}/4)}{2}$$

$$= \frac{l_{+j+}}{4} - \frac{l_{+++}}{8} = D_j$$

$$\frac{S_{1k} + S_{2k}}{2} = \frac{l_{++k}}{4} - \frac{l_{+++}}{8} = S_k$$

and

$$\frac{(DS)_{1jk} + (DS)_{2jk}}{2} = \frac{l_{+jk}}{2} - \frac{l_{+j+} + l_{++k}}{4} + \frac{l_{+++}}{8} = (DS)_{jk}.$$

That is, the averages of the two sets of two-way-table parameters generate new parameters which are similar to the collapsed table parameters.

The differences or "contrasts" of the two sets of two-way-table parameters yield parameters for characterizing between clinic effects. These are

$$\frac{(M_1 - M_2)}{2} = \frac{l_{1++}/4 - l_{2++}/4}{2} = \frac{l_{1++}}{8} - \frac{l_{+++} - l_{1++}}{8}$$

$$= \frac{l_{1++}}{4} - \frac{l_{+++}}{8} = C_1 = -C_2$$

$$\frac{D_{1j} - D_{2j}}{2} = \frac{(l_{1j+}/2 - l_{1++}/4) - (l_{2j+}/2 - l_{2++}/4)}{2}$$

$$= \frac{(l_{1j+}/2 - l_{1++}/4) - [(l_{+j+} - l_{1j+})/2 - (l_{+++} - l_{1++})/4]}{2}$$

$$= \frac{l_{1j+}}{2} - \frac{l_{1++} + l_{+j+}}{4} + \frac{l_{+++}}{8} = (CD)_{1j} = -(CD)_{2j}$$

$$\frac{S_{1k} - S_{2k}}{2} = \frac{l_{1+k}}{2} - \frac{l_{1++} + l_{++k}}{4} + \frac{l_{+++}}{8}$$

$$= (CS)_{1k} = -(CS)_{2k}$$

and

$$\frac{(DS)_{1jk} - (DS)_{2jk}}{2} = l_{1jk} - \frac{l_{1j+} + l_{1+k} + l_{+jk}}{2}$$

$$+ \frac{l_{1++} + l_{+j+} + l_{++k}}{4} - \frac{l_{+++}}{8}$$

$$= (CDS)_{1jk} = -(CDS)_{2jk}.$$

The combined log-linear model is given by

$$l_{ijk} = \ln(m_{ijk}) = M + C_i + D_j + S_k$$
$$+ (CD)_{ij} + (CS)_{ik} + (DS)_{jk} + (CDS)_{ijk} \qquad (4.47)$$

where the parameters are defined and interpreted in Table 4.9. Notice that the standard identifiability conditions are implied by the parameter definitions:

$$\sum_i C_i = \sum_j D_j = \sum_k S_k = \sum_i (CD)_{ij} = \sum_j (CD)_{ij} = \sum_i (CS)_{ik}$$

$$= \sum_k (CS)_{ik} = \sum_j (DS)_{jk} = \sum_k (DS)_{jk} = \sum_i (CD)_{ijk}$$

$$= \sum_j (CDS)_{ijk} = \sum_k (CDS)_{ijk} = 0.$$

The overall mean (M) is the logarithm of the geometric mean of the cell counts. It corresponds to the constraint that $\Sigma_i \Sigma_j \Sigma_k m_{ijk} = 715 = n$ or $\Sigma_i \Sigma_j \Sigma_k p_{ijk} = 1$. The main effects are the log-geometric means of the odds of a given category of the corresponding variable, where the mean is taken across all levels of the remaining variables. This is because, for example, $C_1 = l_{1++}/4 - l_{+++}/8 = (l_{1++} - l_{2++})/8$. The pairwise interactions $(CD)_{ij}$, $(CS)_{ik}$, $(DS)_{jk}$ are the log-geometric means of the odds ratios where the mean is across the levels of the third variable. Finally, the triple interaction $(CDS)_{ijk}$ is the log ratio of the odds ratios between two variables, where the ratio is across the levels of the third variable. The parameters, the results for stratified tables, and various interpretations are shown in Table 4.9.

Commonly, the two-variable interaction term is called the *adjusted association*. The adjustment is with respect to the third variable. Alternatively, the interaction is referred to as the *partial association* between two variables where the effect of the third variable has been "partialed out." It may also be referred as the first-order interaction. The triple interaction parameter is also referred to as the second-order interaction. When associations are discussed it is simply the interaction.

We can apply these general results to observed data in the usual manner. The m.l.e.'s for the saturated model may be obtained by replacing the m_{ijk} with y_{ijk}, the observed counts. For the data in Table 4.8 results are

$$\hat{M} = \frac{\ln [3(176)(4)(293)(17)(197)(2)(23)]}{8} = 3.161$$

$$\hat{C}_1 = \frac{\ln [3(176)(4)(293)/(17)(197)(2)(23)]}{8} = 0.174$$

$$\hat{D}_1 = \frac{\ln [3(176)(17)(197)/(4)(293)(2)(23)]}{8} = 0.436$$

Table 4.9 Parameterization of Log-Linear Model for 2 × 2 × 2 Table, Standard Identifiability Conditions at Cell (1,1,1)

Parameter (1)	Stratified Representation (2)	Odds Interpretation (3)	Log Count Representation (4)	Log Ratio Representation (5)
m	$\frac{m_1 + m_2}{2}$	Geometric mean of cell counts	$\frac{l_{+++}}{8}$	$[\ln (m_{111}m_{112}m_{121}m_{122}m_{211}m_{212}m_{221}m_{222})]/8$
C_1	$\frac{m_1 - m_2}{2}$	Average odds of clinic 1	$\frac{l_{1++}}{4} - \frac{l_{+++}}{8}$	$\left[\ln\left(\frac{m_{111}m_{112}m_{121}m_{122}}{m_{211}m_{212}m_{221}m_{222}}\right)\right]/8$
D_1	$D_{11} + D_{21}$	Average odds of less care	$\frac{l_{+1+}}{4} - \frac{l_{+++}}{8}$	$\left[\ln\left(\frac{m_{111}m_{112}m_{211}m_{212}}{m_{121}m_{122}m_{221}m_{222}}\right)\right]/8$
S_1	$\frac{S_{11} + S_{21}}{2}$	Average odds of infant dying	$\frac{l_{++1}}{4} - \frac{l_{+++}}{8}$	$\left[\ln\left(\frac{m_{111}m_{121}m_{211}m_{221}}{m_{112}m_{122}m_{212}m_{222}}\right)\right]/8$
$(CD)_{11}$	$\frac{D_{11} - D_{21}}{2}$	Average odds ratio: clinic and care	$\frac{l_{11+}}{2} - \frac{l_{1++} + l_{+1+}}{4} + \frac{l_{+++}}{8}$	$\left[\ln\left(\frac{m_{111}m_{221}}{m_{121}m_{211}} \cdot \frac{m_{112}m_{222}}{m_{122}m_{212}}\right)\right]/8$
$(CS)_{11}$	$\frac{S_{11} - S_{21}}{2}$	Average odds ratio: clinic and survival	$\frac{l_{1+1}}{2} - \frac{l_{1++} + l_{++1}}{4} + \frac{l_{+++}}{8}$	$\left[\ln\left(\frac{m_{111}m_{212}}{m_{112}m_{211}} \cdot \frac{m_{121}m_{222}}{m_{122}m_{221}}\right)\right]/8$
$(DS)_{11}$	$\frac{DS_{111} - DS_{211}}{2}$	Average odds ratio: care and survival	$\frac{l_{+11}}{2} - \frac{l_{+1+} + l_{++1}}{4} + \frac{l_{+++}}{8}$	$\left[\ln\left(\frac{m_{111}m_{122}}{m_{121}m_{112}} \cdot \frac{m_{211}m_{222}}{m_{221}m_{212}}\right)\right]/8$
$(CDS)_{111}$	$DS_{111} - DS_{211}$	Ratio of odds ratios	$l_{111} - \frac{l_{11+} + l_{1+1} + l_{+11}}{2} + \frac{l_{1++} + l_{+1+} + l_{++1}}{4} - \frac{l_{+++}}{8}$	$\left[\ln\left(\frac{m_{111}m_{221}}{m_{121}m_{211}} \cdot \frac{m_{122}m_{212}}{m_{112}m_{222}}\right)\right]/8$

$$\hat{S}_1 = \frac{\ln [3(4)(17)(2)/(176)(293)(197)(23)]}{8} = 1.657$$

$$(\hat{CD})_{11} = \frac{\ln \{[3(2)/(4)(17)][176(23)/(293)(197)]\}}{8} = -0.636$$

$$(\hat{CS})_{11} = \frac{\ln \{[3(197)/(17)(176)][4(23)/(2)(293)]\}}{8} = -0.434$$

$$(\hat{DS})_{11} = \frac{\ln \{[3(293)/(4)(176)][17(23)/(2)(197)]\}}{8} = 0.027$$

$$(\widehat{CDS})_{111} = \frac{\ln \{[3(2)/(4)(17)][176(23)/(293)(197)]\}}{8} = 0.029.$$

The large-sample standard error for each of these parameters is a straightforward generalization of the standard error of the log odds ratio in 2 × 2 tables:

$$s = \left[\frac{\Sigma_i \Sigma_j \Sigma_k (1/y_{ijk})}{8} \right]^{1/2} = 0.137.$$

On comparing the parameter estimates to s, neither $(CDS)_{111}$ nor $(DS)_{11}$ appear to be significantly different from zero. However, $(CD)_{11}$ and $(CS)_{11}$ are more than three and four times s, respectively. Hence they are statistically significant. These observations lead to the following conclusions:

1. The association between any two variables does not depend on the level of the third. For example, the association between duration of care and survival does not depend on which clinic is considered. This is because H_0: $CDS = 0$ is accepted.
2. The association between duration and survival is not significant. This follows from accepting H_0: $DS = 0$ or the variables duration of care and survival are independent.
3. There is a "negative" association between clinic and duration of care. Here H_0: $CD = 0$ is rejected and we conclude that the variables clinic and duration of care are dependent, after adjusting for survival.

Moreover, since $\hat{CD}$ is the product of the survival specific odds ratios, the corresponding "adjusted odds ratio" is exp [4(−0.636)] = 0.079. That is, mothers in clinic 1 are much less likely to receive less than 1 month of care than mothers in clinic 2. The odds ratio is "adjusted" for survival.

4. There is a "negative" association between clinic and survival. That is, H_0: $CS = 0$ is rejected and we can say the variables "clinic" and "survival" are dependent.

Again using the adjusted odds ratio, we conclude the odds of an infant surviving are significantly higher in clinic 1. The marginal association between duration of care and survival now appears to be spurious in the sense that once the association is adjusted for "clinic," it vanishes. The conditions under which this occurs are discussed in detail in Section 4.6.

It might be argued that including the interaction term (CDS) in the model was unnecessary since this was not a study hypothesis a priori. A more direct, and possibly more powerful, approach would be to fit a simpler model and test the DS term directly. One such simpler model is

$$l_{ijk} = \ln (m_{ijk}) = M + C_i + D_j + S_k + (CD)_{ij} + (CS)_{ik} + (DS)_{jk}. \quad (4.49)$$

The parameters and m_{ijk} can be estimated and a chi-square statistic applied to test

$$H_0: \quad (CDS)_{ijk} = 0. \quad (4.50)$$

If this hypothesis is accepted, then the even simpler model

$$l_{ijk} = \ln (m_{ijk}) = M + C_i + D_j + S_k + (CD)_{ij} + (CS)_{ik} \quad (4.51)$$

can be fit. Estimated expected values from this model can be used to test

$$H_0: \quad (DS)_{jk} = (CDS)_{ijk} = 0. \quad (4.52)$$

If likelihood ratio chi-square tests have been used, the difference between that in (4.52) and (4.50) will yield a likelihood ratio test of

$$H_0: \quad (DS)_{jk} = 0. \tag{4.53}$$

The test is sometimes referred to as a "direct test" of (4.53), but more correctly, it is conditional on accepting the no-interaction hypothesis (4.50). Alternatively, it is conditional on the data fitting the model (4.49).

The only difficulty with this approach is that the parameters in (4.49) cannot be estimated directly. Instead, some type of iterative procedure such as iterative proportional fitting (IPF) (Bishop et al. 1975), Newton–Raphson (Haberman 1978), or iterated weighted least squares (Nelder and Wedderburn 1972) must be employed. The IPF approach has been implemented in the software BMDP4F (Dixon 1981). A brief summary of the approach is found in Appendix 4A.3.

Applying these algorithms, all three models [(4.47), (4.49), and (4.51)] can be estimated and the corresponding likelihood ratio test computed. The results are shown in Table 4.10. Except for rounding error, the parameter estimates for the saturated model (4.47) are identi-

Table 4.10 Maximum Likelihood Estimates and Likelihood Ratio Tests for Three Models of the Association Between Survival, Duration of Care, and Clinic (Table 4.8)

Parameter	Estimates[a]		
Model (equation)	(4.47)	(4.49)	(4.51)
M	3.160	3.148	3.154
C_1	0.174	0.181	0.169
D_1	0.436	0.435	0.410
S_1	−1.657	−1.671	−1.663
$(CD)_{11}$	−0.636	−0.662	−0.663
$(CS)_{11}$	−0.434	−0.425	−0.439
$(DS)_{11}$	0.027	0.027	0
$(CDS)_{11}$	0.029	0	0
G^2	0	0.04	0.08
df	0	1	2

[a]Zero denotes hypothesis.
Source: Computations using BMDP4F (Dixon 1981).

cal to what was computed by hand. The likelihood ratio tests for the simpler models (4.49) and (4.51) are 0.04 and 0.08, respectively. The degrees of freedom can be computed either by counting the number of parameters set equal to zero (1 and 2), or by subtracting the number of estimated parameters from the number of cells: (4.49): $8 - 7 = 1$, (4.51): $8 - 7 = 2$. The conditional test of (4.53) is obtained by comparing the fit of the simpler model to that of the more complex model: $0.08 - 0.04 = 0.04$. Hence we conclude that $(DS)_{jk} = 0$, as before.

In summary, the general form of the log-linear model in three dimensions defined by $A = 1, 2, \ldots, I$, $B = 1, 2, \ldots, J$, and $C = 1, 2, \ldots, K$ is

$$l_{ijk} = \ln (m_{ijk}) = M + A_i + B_j + C_k + (AB)_{ij} + (AC)_{ik} + (BC)_{jk} + (ABC)_{ijk}, \tag{4.54}$$

where

$$M = \frac{\sum_{i=1}^{I} \sum_{j=1}^{J} \sum_{k=1}^{K} l_{ijk}}{IJK} \qquad A_i = \frac{l_{i++}}{JK} - \frac{l_{+++}}{IJK}$$

$$B_j = \frac{l_{+j+}}{IK} - \frac{l_{+++}}{IJK} \qquad C_k = \frac{l_{++k}}{IJ} - \frac{l_{+++}}{IJK}$$

$$(AB)_{ij} = \frac{l_{ij+}}{K} - \frac{l_{i++}}{JK} - \frac{l_{+j+}}{IK} + \frac{l_{+++}}{IJK}$$

$$(AC)_{ik} = \frac{l_{i+k}}{J} - \frac{l_{i++}}{JK} - \frac{l_{++k}}{IJ} + \frac{l_{+++}}{IJK}$$

$$(BC)_{jk} = \frac{l_{+jk}}{I} - \frac{l_{+j+}}{IK} - \frac{l_{++k}}{IJ} + \frac{l_{+++}}{IJK}$$

$$(ABC)_{ijk} = l_{ijk} - \frac{l_{ij+}}{K} - \frac{l_{i+k}}{J} - \frac{l_{+jk}}{I} + \frac{l_{i++}}{JK} + \frac{l_{+j+}}{IK} + \frac{l_{++k}}{IJ} - \frac{l_{+++}}{IJK}.$$

Notice that this model implies the usual identifiability conditions:

$$\sum_i A_i = \sum_j B_j = \sum_k C_k = \sum_i (AB)_{ij} = \sum_j (AB)_{ij} = \sum_i (AC)_{ik}$$

$$= \sum_k (AC)_{ik} = \sum_j (BC)_{jk} = \sum_k (BC)_{jk} = \sum_i (ABC)_{ijk}$$

$$= \sum_j (ABC)_{ijk} = \sum_k (ABC)_{ijk} = 0.$$

Thus there are IJK parameters describing IJK cells, so the model is termed a *saturated model.* Simpler models are formed by deleting terms from the model. In the case of the model with no three-way interaction, $(ABC)_{ijk} = 0$, it must be estimated by iterative methods.

It is usual practice to include all components of an interaction term whenever it is included. For example, the inclusion of $(AB)_{ij}$ implies the inclusion of M, A_i, B_j, as well. This is known as the *hierarchy principle.* Later, we examine some variations that will simplify the interpretation of complex interaction terms, but for the moment we restrict our attention to *hierarchical models.* Thus the model of no three-way interaction can be written

$$l_{ijk} = \ln(m_{ijk}) : [AB, AC, BC]. \tag{4.55}$$

where both indices and lower-order terms are implied.

As noted in Section 4.3, log-linear models are frequently written as *u-models.* The u-model notation for the saturated model is

$$\ln(m_{ijk}) = u + u_{1(i)} + u_{2(j)} + u_{3(k)} + u_{12(ij)} + u_{13(ik)} + u_{23(jk)} + u_{123(ijk)}. \tag{4.56}$$

4.5 MODEL SELECTION

The log-linear model for three-dimensional tables is particularly useful when no single variable is designated as the response. In this situation specific hypotheses may be of less interest than characterization of the complete structure of the data. As an example, the data in Table 4.11 show the data in Table 4.7 subdivided by village. The question of interest is: What is the minimum number of parameters required to characterize the data? By "characterize the data," we mean that the model has to "fit the data." Here we have reversed the usual terminol-

Table 4.11 Sex Distribution of the Yanomamö by Lineage and Village

Sex (S)	Lineage (L)		Villages (V): Bisaasi-teri, Lower $(i, j, 1)$	Bisaasi-teri, Upper $(i, j, 2)$	Monou-teri $(i, j, 3)$	Patanowä-teri $(i, j, 4)$	Total $(i, j, +)$
			Observed Counts				
Male	Sha	$(1, 1, k)$	9	11	11	39	70
	Hor	$(1, 2, k)$	1	8	15	58	82
	Shamatar	$(1, 3, k)$	3	3	3	2	11
	Other	$(1, 4, k)$	12	15	5	23	55
Female	Sha	$(2, 1, k)$	7	14	9	29	59
	Hor	$(2, 2, k)$	2	5	8	33	48
	Shamatar	$(2, 3, k)$	9	7	9	6	31
	Other	$(2, 4, k)$	8	22	6	22	58
Total			51	85	66	212	414
			Fitted Counts				
Male	Sha		8.682	13.566	10.853	36.899	70
	Hor		1.892	8.200	14.508	57.400	82
	Shamatar		3.143	2.619	3.143	2.095	11
	Other		9.735	18.009	5.354	21.903	55
Female	Sha		7.318	11.434	9.147	31.101	59
	Hor		1.108	4.800	8.492	33.600	48
	Shamatar		8.857	7.381	8.857	5.905	31
	Other		10.265	18.991	5.646	23.097	58
Total			51	85	66	212	414

Source: Chagnon (1977).

ogy to emphasize that we are fitting a model to the data, whereas in theory we test the model with a data set.

Suppose that Y_{ijk} denotes the count in the (i, j, k) cell corresponding to the ith sex, jth lineage, and kth village, $i = 1, 2, j = 1, 2, 3, 4, k = 1, 2, 3, 4$. We assume that Y_{ijk} follows a multinomial distribution for purposes of estimation and for testing hypotheses. The saturated model is

$$l_{ijk} = M + S_i + L_j + V_k + (SL)_{ij} + (SV)_{ik} + (LV)_{jk} + (SLV)_{ijk} \quad (4.57)$$

where the effects are defined as in (4.54). If we restrict our attention to hierarchical models, there are 17 possible models that can be considered. These are enumerated together with the corresponding G^2 statistics in Table 4.12.

Here the notation indicates the most complex terms and implies the lower-order relatives. The degrees of freedom correspond to the number of parameters set equal to zero. Thus [*SL*, *LV*] corresponds to

$$l_{ijk} = M + S_i + L_j + V_k + (SL)_{ij} + (LV)_{jk}, \quad (4.58)$$

Table 4.12 All-Possible-Models Analysis of Three-Variable Table

Model	df	G^2	Selected Model	Δdf	ΔG^2	Deleted Terms
SL,SV,LV	9	2.82	1	—	—	*LSV*
SL,SV	18	69.57		9	66.75	*LV*
SL,LV	12	4.78	2	3	1.96	*SV*
SV,LV	12	18.36		3	15.54	*LV*
SL,V	21	74.76		9	69.98	*LV*
SL,L	21	88.34		9	83.56	[a]
S,LV	15	23.55	3	3	18.77	*SL*
SV	24	148.51				
SL	24	213.71				
LV	16	24.72	4	1	1.17	*S*
S,V,L	24	93.52		9	69.97	*LV*
S,V	27	153.70				
S,L	27	232.48				
V,L	25	94.69	5	9	69.97	
S	30	292.66				
V	28	154.87				
L	28	233.65				

[a]Nonhierarchical model given previous step.

where $(SV)_{ik} = (SLV)_{ijk} = 0$. Since there are three distinct values to $(SV)_{ik}$ and nine distinct values to $(SLV)_{ijk}$ under the usual identifiability conditions, df = 3 + 9 = 12. Alternatively, we can count the number of estimated parameters and subtract these from the number of cells:

$$\text{df} = 2(4)(4) - (1 + 1 + 3 + 3 + 3 + 9) = 32 - 20 = 12.$$

We can now use the tabulation of the test statistics to identify an appropriate model. The process is to eliminate the most complex terms in descending order of statistical significance. This is analogous to "backwards elimination" in multiple regression (Kleinbaum and Kupper 1978, Draper and Smith 1981). The first step is to examine the model of no three-way interaction, denoted [*SL*, *SV*, *LV*]. The corresponding likelihood ratio chi-square is 2.82 with df = 9; that is, *SLV* is not significantly different from zero.

There are three possible lower-order relatives of the no-three-way-interaction model. These are obtained by deleting each pairwise effect in turn. Of these three, the model [*SL*, *LV*] fits quite well. Moreover, the change in chi-square is quite small: 4.78 − 2.82 = 1.96. This indicates that the *SV* term is nonsignificant (df = 12 − 9 = 3), given no three-way interaction.

The third step is to compute the lower-order relatives of [*SL*, *LV*]: [*SL*, *V*] and [*S*, *LV*]. The former does not fit at all well, but the latter has a marginal fit ($G^2 = 23.55$, df = 15, *p*-value = 0.07). However, when we compare [*S*, *LV*] to the previously selected model, it is evident the *SL* term is significant ($\Delta G^2 = 23.55 - 4.78 = 18.77$, df = 15 − 12 = 3). Hence *SL* should be retained. Notice that the model [*SV*, *L*] is not considered since it is not a lower-order relative of [*SL*, *LV*]. The backwards elimination algorithm therefore selects the model [*SL*, *LV*] or (4.58). Since all other models have been examined, we can check to see whether any other model is plausible. The only candidate is [*LV*], but this is premised on both *SL* and *SV* being zero, yet *SL* appears to be significant. A direct test can be obtained by comparing the fit of [*LV*] to [*SL*, *SV*, *LV*]: 24.72 − 2.82 = 21.90, df = 16 − 9 = 7, which exceeds the usual critical levels. Hence we reject H_0: $SL = SV = 0$.

The backwards elimination of parameters in a hierarchical fashion yields a tentative model for characterizing the data. Before interpreting the model we should attempt to validate our assumptions. One method of validation is termed "residual analysis," which is discussed by a number of authors, including Bishop et al. (1975, pp. 136–141) and Haberman (1978, pp. 17–21 and 77–79). The general approach is to compare the observed to estimated expected values and assess whether the "residuals" are normally distributed. It is important to recognize that there are no more distinct residuals to be considered than degrees of freedom for the model.

There are two common methods for standardizing or normalizing the residuals, adjusted residuals (see Section 3.4.2), and Freeman–Tukey deviates:

$$z_k = \frac{y_{ijk} - \hat{m}_{ijk}}{\sqrt{\hat{m}_{ijk}}} \tag{4.59}$$

$$T_k = \sqrt{y_{ijk}} + \sqrt{(y_{ijk} + 1)} - \sqrt{(4\hat{m}_{ijk} + 1)} \tag{4.60}$$

where $k = 1, 2, \ldots, s$; s = degrees of freedom for the model. Either can be used in the analysis, but we will restrict attention to the former. One approach is to plot the standardized residuals against their expected values under the assumption of normality. If any residual is too large, it is termed an *outlier* and investigated in greater detail. If the overall pattern deviates sharply from what is expected, alternative models should be considered. One method of determining what is expected is the *rankit plot* discussed by Haberman (1978, 1979). It can be illustrated using the Yanomamö data.

Two models of some interest based on the backwards elimination procedure are *SL, LV* and *S, LV,* with 12 and 15 degrees of freedom, respectively. In the former, *SLV* and *SV* have been set equal to zero. This is equivalent to saying that the *SV* expected values are unconstrained but the *SL* and *LV* marginal tables are fixed. The rationale behind the statement is discussed more fully in Section 4.7. Thus the *SL* and *LV* marginal tables are fixed. In columns (1) and (2) of Table 4.13, the observed counts and estimated expected counts for the male by village by lineage subtable are shown. Since the *SL* table is fixed, the female values are redundant. Similarly, the Patanowa by lineage counts are made redundant by the *LV* margin being fixed. The specifi-

Table 4.13 Rankit Analysis of *SL,LV* and *S,LV* Log-Linear Models of Yanomama (Table 4.11): Males Only

	Model:	Observed	*SL,LV*					*S,LV*	
			Expected	Adjusted Residual	Rank	$\frac{3k-1}{3s+1}$	Rankit	Adjusted Residual	Rankit
		y	m	$y - m$	k				
Lineage	Village	(1)	(2)	(3)	(4)	(5)	(6)	(7)	(8)
Sha	Lower	9	8.68	0.11	8	0.62	0.33	0.20	0.00
	Upper	11	13.57	−0.67	2	0.14	−1.10	−0.16	−0.33
	Monou	11	10.85	0.04	7	0.54	0.12	0.14	−0.12
	Patanowa	39	—	—	—	—	—	0.53	0.74
Hor	Lower	1	1.89	−0.65	3	0.22	−0.79	−0.46	0.51
	Upper	8	8.20	−0.07	6	0.46	−0.10	0.44	0.36
	Monou	15	14.51	0.13	9	0.70	0.54	0.83	0.93
	Patanowa	58	—	—	—	—	—	1.46	1.68
Shamatar	Lower	3	3.14	−0.08	4.5	0.34	−0.42	−1.32	−1.43
	Upper	3	2.62	0.24	11	0.86	1.10	−0.99	−0.71
	Monou	3	3.14	−0.08	4.5	0.34	−0.42	−1.32	−1.43
	Patanowa	2	—	—	—	—	—	1.08	1.28
Other	Lower	12	9.74	0.73	12	0.95	1.59	0.45	0.51
	Upper	15	18.01	−0.71	1	0.05	−1.61	−1.02	−0.94
	Monou	5	5.35	0.15	10	0.78	0.79	0.33	0.16
	Patanowa	—	—	—	—	—	—	—	—

cation of the female and Patanowa counts being redundant is completely arbitrary, so the redundant cells should also be inspected for possible outliers, as was done in Section 3.4.2.

In column (3) the adjusted residuals, z_h, $h = 1, \ldots, 12$ have been computed and the corresponding ranks are shown in column (4). For example, $z_2 = (11 - 13.57)/\sqrt{13.57} = -0.70$. This is the second smallest adjusted residual, so it receives rank 2. Notice that z_7 and z_9 are tied, so they receive the midrank value of 4.5.

The next step is to standardize the ranks:

$$r_h = \frac{3k - 1}{3s + 1} \tag{4.61}$$

where k is the rank of the hth residual. For example, $r_3 = [3(7) - 1]/[3(12) + 1] = 20/37 = 0.54$. If the adjusted residuals follow a standard normal distribution, this should correspond to the area of the appropriate percentile of the standard normal distribution. This is termed the *rankit*. For $h = 3$ the rankit is 0.12 since $P(z \leq 0.54) = 0.12$. The remaining rankits are shown in column (6).

The rankit plot is constructed by plotting the adjusted residuals against the corresponding rankits, as in Figure 4.1. If the residuals are exactly normal, they will fall on the 45° line through the origin (slope of 1). If they are too large, they will fall along a line with a slope greater than 1. In the present example the circled points tend to fall along a line with a slope of less than 1. This suggests that too "good" a fit has been achieved since there are too few large residuals. The very large p-value (0.9649) for the fit of this model reinforces this view of too good a fit.

The residual analysis indicates that a lower-order relative of [*SL*, *LV*] should be considered. The adjusted residuals and rankits for [*S*, *LV*] are shown in columns (7) and (8) in Table 4.13. Note that the female cells are still redundant, but since *SL* has been eliminated, only the Other–Patanowa cell is also redundant. Thus we have 15 points to plot in Figure 4.2. The rankit plot for [*S*, *LV*] (as well as [*LV*]) is much closer to the normal line, which suggests that this model is somewhat better than [*SL*, *LV*] in the sense of not being "overfitted." A plot of [*S*, *L*, *V*] would indicate the presence of several large residuals, indicating the clear significance of the *LV* parameters.

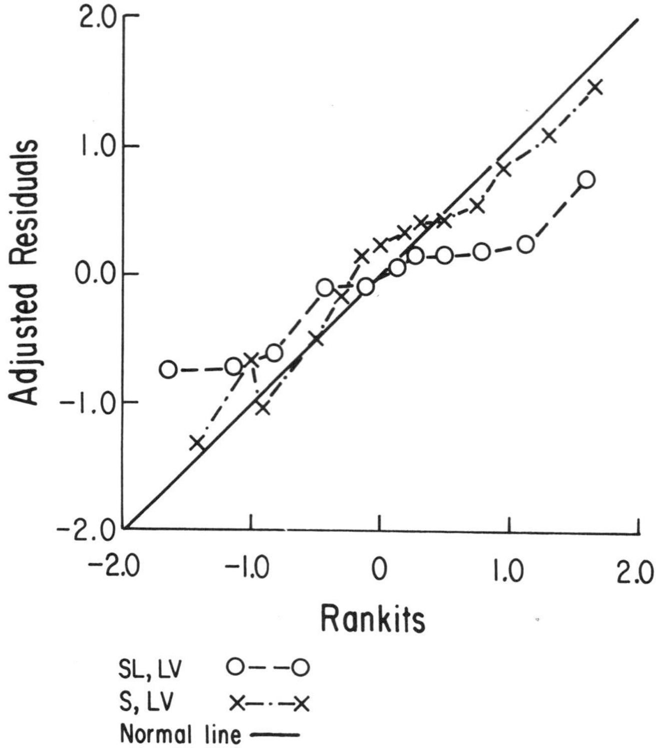

Figure 4.1 Rankit plots for adjusted residuals of *SL,LV* and *S,LV* Yanomama data.

At this point the analyst is in something of a dilemma. The model selection indicates that the SL parameters are "statistically significant," but the residual analysis suggests that the model $[SL, LV]$ overfits the data. There is no clear-cut resolution to this dilemma. The most reasonable approach is to conclude a dependence between lineage and village but retain a healthy skepticism about the relationship between sex and lineage.

In summary, the process of model selection for three variables is:

1. Compute the tests of fit for the 17 possible models.
2. Eliminate parameter groups in a hierarchical fashion beginning with the highest-order interactions.
3. Perform a residual analysis of the selected models.

4.6 COLLAPSIBILITY

At the beginning of this chapter it was observed that three-way tables are of interest when an investigator wishes to determine whether an association between two variables depends on a third. One approach to this issue has been developed in terms of the test for three-way interaction. For the two-factor, one-response analysis this was the test for no interaction between the factors. Similarly, in the three-response analysis for independence, it corresponded to hypotheses of the form

$$H_0: \ (ABC)_{ijk} = 0 \qquad i = 1, \ldots, I; \ \ j = 1, \ldots, J; \\ k = 1, \ldots, K. \qquad (4.62)$$

Once we establish that there is no interaction, simpler models are examined to characterize the data more simply. A related question is whether the complete three-dimensional table is necessary for summarizing the data, or can two-dimensional tables by validly reported? That is, when can the study association between two variables be validly collapsed across the levels of the third variable?

As an example, consider Bartlett's data (Table 4.1) as if they are of special interest to determine the appropriate length of cutting. For the moment suppose that the time of planting is treated as the third or stratification variable. Sometimes this third variable is termed a *confounding* or *extraneous variable*. Is it necessary to report the full three-dimensional table as at the top of Table 4.1, or can the simpler 2 × 2 table of length by survival be validly reported? Here the index of association is the simple difference in the survival rates.

The average survival rate in the three-dimensional table is

$$\overline{D} = \frac{(156/240 - 107/240) + (84/240 - 31/240)}{2}$$

$$= \frac{(0.2042 + 0.2208)}{2} = 0.212.$$

The value in the collapsed table is $D_+ = (240/480 - 138/480) = 0.212 = \overline{D}$. Clearly, the simple difference is collapsible with respect to time of planting. Also, it can be shown that the simple difference in survival rates for time of planting can be collapsed across length of cutting.

Does this relationship hold for the relative risks and odds ratios? First, consider the average and marginal relative risks:

$$\overline{RR} = \left[\left(\frac{156/240}{107/240}\right)\left(\frac{84/240}{31/240}\right)\right]^{1/2}$$

$$= [(1.4579)(2.7097)]^{1/2} = 1.99$$

$$RR_+ = \frac{240/240}{138/240} = 1.74.$$

In this case the average does not equal the marginal value, so the relative risk is not collapsible over time of planting. Notice that the simple differences are nearly constant but the relative risks are quite different.

Next consider the odds ratios:

$$\overline{OR} = \left[\left(\frac{156/84}{107/133}\right)\left(\frac{84/156}{31/209}\right)\right]^{1/2} = 2.89$$

$$OR_+ = \frac{240/240}{138/342} = 2.48.$$

Again the odds ratio is not collapsible with respect to time of planting. It will turn out that the relative risk is not collapsible because of the "interaction" between time of planting and length of cutting. Even if $\overline{\text{RR}} = \text{RR}_+$, collapsing the table is misleading because the association depends on which level of the third variable is considered. However, for the odds ratio the reason is more subtle. It turns out that given survival, time and length are related. Also given length of cutting, time and survival are related. These two *conditional associations* lead to the odds ratio not being collapsible, at least for $2 \times 2 \times 2$ tables.

The data on the Yanomamö are an example of when odds ratios, given an appropriate model, are collapsible. Suppose that we fit the model $[SV, LV]$ (4.58), that is, the model of no three-way interaction ($SLV = 0$) and no conditional association between sex and lineage ($SL = 0$). The resulting estimated expected values are the second entries in Table 4.11. Notice that the village and sex by lineage totals are identical for the observed and fitted values. (We use the term "fitted values" instead of the more awkward phrase "estimated expected values under the model.") It can be verified that the lineage by village observed and fitted totals are also identical. This is a general property of log-linear models which will be examined in more detail in Section 4.7.

The collapsed odds ratios relating the sex distribution by lineage are Sha-Hor: $(70/59)/(82/48) = 0.696$; Sha-Shamatar: $(70/59)/(11/31) = 3.353$; Sha-Other: $(70/59)/(55/58) = 1.255$. Now consider the same comparison in Lower Bisaasi-Teri: Sha-Hor: $(8.682/7.318)/(1.892/1.108) = 0.695$; Sha-Shamatar: $(8.682/7.318)/(3.143/8.857) = 3.353$; Sha-Other: $(8.682/7.318)/(9.735/10.265) = 1.2510$. These are identical to those in the collapsed table except for rounding error. Moreover, the same result holds for each of the other villages.

This illustrates two points about odds ratios:

1. When $SL = 0$, then SV is collapsible with respect to L if $SLV = 0$.
2. When $SLV = 0$, then the two-variable associations are constant over the levels of the third variable.

It turns out the LV odds ratios under this model are also collapsible with respect to S. Finally, the odds ratios for SV are not collapsible with respect to lineage because while $SLV = 0$, $SL \neq 0$ *and* $SV \neq 0$.

This general issue of collapsibility was first pointed out by Simpson (1951). Necessary and sufficient conditions for collapsibility of the odds ratio in $2 \times 2 \times 2$ tables are given in Bishop et al. (1975, pp. 39–40). The generalization to $I \times J \times K$ and higher-dimensional tables is found in Whittemore (1978). A geometric proof of particular interest in $I \times 2 \times 2$ tables is found in Shapiro (1982). He also sketches an equivalent proof for the relative risk. Special conditions for the relative risks are also found in Kleinbaum et al. (1982). In all cases, interest is primarily in the situation where the index of association is constant over the levels of the third variable, that is, no three-variable interaction. We will use Shapiro's notation to prove the necessary and sufficient conditions as well as some useful corollaries.

Suppose that we are interested in the association between the variables T and R with levels $j = 1, \ldots, J$ and $k = 1, \ldots, K$, respectively. Let Q_{ih} denote an index of association at level i of a third variable S, $i = 1, \ldots, I$ where $h = 1, \ldots, JK$. Notice there are at most $I(J - 1)(K - 1)$ distinct values of Q_{ih} corresponding to the degrees of freedom in the $I \times J \times K$ tables. Let Q_{+h} denote the marginal association between T and R. By "no three-variable interaction," we mean $Q_{1h} = Q_{2h} = \cdots = Q_{ih} = Q_h$ for $k = 1, \ldots, JK$. The definition of collapsibility is:

Definition 4.1

In the absence of three-variable interaction, Q_{ih} is said to be *collapsible* with respect to S (with I levels) when $Q_h = Q_{+h}$. The collapsibility theorems will be proved for $J = K = 2$, but as noted by Shapiro (1982), they readily generalize.

Let $P(S = i, T = j, R = k) = p_{ijk} > 0$ and assume that the total sample size is n and $E(Y_{ijk}) = np_{ijk} = m_{ijk}$, where $\Sigma_i \Sigma_j \Sigma_k m_{ijk} = n$. Let the conditional probabilities be given by

$$q_{ij} = P(R = 1 | S = i, T = j) = \frac{m_{ij1}}{m_{ij+}}$$
$$q_{+j} = P(R = 1 | T = j) = \frac{m_{+j1}}{m_{+j+}}. \tag{4.63}$$

Now q_+ can be written as a weighted sum of the q_{ij}:

$$q_{+} = P(R = 1|T = j) = \sum_{i=1}^{I} P(S = i, R = 1|T = j)$$

$$= \sum_{i=1}^{I} P(S = i|T = j)P(R = 1|S = i, T = j) = \sum_{i=1}^{I} v_{ij}q_{ij} \quad (4.64)$$

where $v_{ij} = P(S = i|T = j)$. The conditional and marginal simple differences are $D_i = q_{i1} - q_{i2}$ and $D_{+} = q_{+1} - q_{+2}$.

Theorem 4.1

If $D_1 = D_2 = \cdots = \overline{D}_I = D$, then $\overline{D}_{+} = D$ if and only if

$$\sum_{i=1}^{I} v_{i1}q_{ij} = \sum_{i=1}^{I} v_{i2}q_{ij} \quad \text{for all } j. \quad (4.65)$$

Proof: First note that if (4.65) is true for $j = 1$ it is true for $j = 2$. Rewriting (4.65), we have $\Sigma_{i=1}(v_{i1} - v_{i2})q_{i1} = \Sigma_{i=1}(v_{i1} - v_{i2})(1 - q_{i2})$ $= \Sigma_{i=1}(v_{i1} - v_{i2}) - \Sigma_{i=1}(v_{i1} - v_{i2})q_{i2}$. But $\Sigma_{i=1} v_{ij} = \Sigma_{i=1} P(S = i$ $T = j) = 1$, so $\Sigma_{i=1}(v_{i1} - v_{i2})q_{i1} = 0 \Leftrightarrow \Sigma_{i=1}(v_{i1} - v_{i2})q_{i2} = 0$. Now using the weighted averages (4.64), we have

$$D_{+} = q_{+1} - q_{+2} = \sum_{i=1}^{I} v_{i1}q_{i1} - \sum_{i=1}^{I} v_{i2}q_{i2}$$

$$= \sum_{i=1}^{I} v_{i1}(q_{i1} - q_{i2}) + \sum_{i=1}^{I} (v_{i1} - v_{i2})q_{i2}$$

$$= \sum_{i=1}^{I} v_{i1}\overline{D}_i + \sum_{i=1}^{I} (v_{i1} - v_{i2})q_{i2} = D + \sum_{i=1}^{I} (v_{i1} - v_{i2})q_{i2}.$$

Hence $D_{+} = \overline{D}$ if and only if $\Sigma_{i=1}(v_{i1} - v_{i2})q_{i2} = 0$, as was to be shown.

This has two interesting corollaries:

Corollary 4.1.1

If $D_1 = D_2 = \cdots = D_I = D$ and S and T are independent in the marginal table, then D_i is collapsible with respect to S.

Proof: Left as an exercise.

Corollary 4.1.2

If $D_1 = D_2 = \cdots = \overline{D}_I = D$ and S and R are conditionally independent given T, then D_i is collapsible with respect to S.

Proof: Left as an exercise.

It turns out the same results hold for the relative risk, which is defined as $\mathrm{RR}_i = q_{i1}/q_{i2}$ and $\mathrm{RR}_+ = q_{+1}/q_{+2}$.

Theorem 4.2

If RR_1 and $\mathrm{RR}_2 = \cdots = \mathrm{RR}_I = \overline{\mathrm{RR}}$, then $\mathrm{RR}_+ = \overline{\mathrm{RR}}$ if and only if

$$\sum_{i=1}^{I} v_{i1}q_{ij} = \sum_{i=1}^{I} v_{i2}q_{ij}. \tag{4.65a}$$

Proof: Assume $\Sigma_{i=1}\, v_{i1}q_{i1} = \Sigma_{i=1}\, v_{i2}q_{i1} + E$. As before $\mathrm{RR}_+ = \Sigma_{i=1}\, v_{i=1}q_{i1}/(\Sigma_{i=1}\, v_{i2}q_{i2}) = (\Sigma_{i=1}\, v_{i2}q_{i1} + E)/(\Sigma_{i=1}\, v_{i2}q_{i2})$. Now by assumption $q_{i2} = q_{i1}/\overline{\mathrm{RR}}$ so $\mathrm{RR}_+ = \overline{\mathrm{RR}}(\Sigma_i\, v_{i2}q_{i1} + E)/(\Sigma_{i=1}\, v_{i2}q_{i1})$. Simplifying $\mathrm{RR}_+ = \overline{\mathrm{RR}} + \overline{\mathrm{RR}}(E)/\Sigma_{i=1}\, v_{i2}q_{i1}$. Since $\overline{\mathrm{RR}} > 0$ and $\Sigma_{i=1}\, v_{i2}q_{i1} > 0$ it follows that $\mathrm{RR}_+ = \overline{\mathrm{RR}}$ if and only if $E = 0$ which is true if and only if (4.65) is true, as was to be shown.

The same two corollaries follow immediately:

Corollary 4.2.1

If $\mathrm{RR}_1 = \mathrm{RR}_2 = \cdots = \mathrm{RR}_I = \overline{\mathrm{RR}}$ and S and T are independent in the marginal table, then RR_i is collapsible with respect to S.

Corollary 4.2.2

If $\mathrm{RR}_1 = \mathrm{RR}_2 = \cdots = \mathrm{RR}_I = \overline{\mathrm{RR}}$ and S and R are conditionally independent then RR_i is collapsible with respect to S.

A similar theorem holds for the odds ratio with the important difference that the weights are the conditional probabilities of S given T *and* R. Let the simple odds at $S = i = 1, \ldots, I$ be denoted $O_{ij} = P(R = 1|S = i, T = j)/P(R = 2|S = i, T = j) = m_{ij1}/m_{ij2}$, and the marginal odds be denoted $O_+ = P(R = 1|T = j)/P(R = 2|T = j) = m_{+j1}/m_{+j2}$. As before, the marginal odds can be written as a weighted sum of the conditional odds:

$$O_+ = \sum_{i=1}^{I} w_{ij}O_{ij}, \tag{4.66}$$

where $w_{ij} = P(S = i|T = j, R = 2) = m_{ij2}/m_{+j+}$. The conditional and marginal odds ratios are given by $\mathrm{OR}_i = O_{i1}/O_{i2}$ and $\mathrm{OR}_+ = O_{+1}/O_{+2}$. Using this formulation it follows that

Theorem 4.3
If $OR_1 = OR_2 = \cdots = OR_I = \overline{OR}$, then OR_i is collapsible with respect to S if and only if $\Sigma_{i=1} w_{i1}O_{ij} = \Sigma_{i=1} w_{i2}O_{ij}$, for all $j = 1, 2$.

Proof: Same as for Theorem 4.2 with w_{ij} replacing v_{ij}.

Corollary 4.3.1
If $OR_1 = OR_2 = \cdots = OR_I = \overline{OR}$, then OR_i is collapsible with respect to S if

(i) either S and T are conditionally independent given R
(ii) or S and R are conditionally independent given T.

Proof: Left as an exercise.

The collapsibility theorems have four important ramifications, two of which affect analysis and two of which affect study design and reporting. The first point is that when $I = 2$, the corollaries provide necessary and sufficient conditions for collapsibility. This corresponds to what is known as *Simpson's paradox.*

The second point is that when $I > 2$, some tables may still be collapsible yet there is no convenient statistical test. One approach for odds ratios, suggested by Shapiro (1982) and extended to other indices by Freeman (1983), is to plot the indices with the numerators defining the abscissa values and the numerators defining the ordinates. If there is no three-way interaction, a single line will be defined by the points. If the marginal index falls on this line, the index is collapsible.

For example, in Figure 4.2 the survival rates have been plotted for long and short cuttings. The two 45° lines are nearly collinear, indicating that the simple difference does not depend on the time of planting. In addition, the marginal odds also nearly falls on the same lines; hence it is appropriate to collapse the data for reporting the length of cutting effects.

The third point follows immediately. Frequently, a data set is quite complex. The collapsibility theorems indicate when simpler contingency tables can be validly reported. For example, since the sex-by-lineage and village-by-lineage associations are each collapsible with respect to village and sex, respectively, it is valid to show tables such as the sex-by-lineage contingency table (Table 4.7). Notice that the corresponding likelihood ratio tests of independence are also identical. On the other hand, it is inappropriate to report the sex-by-village table

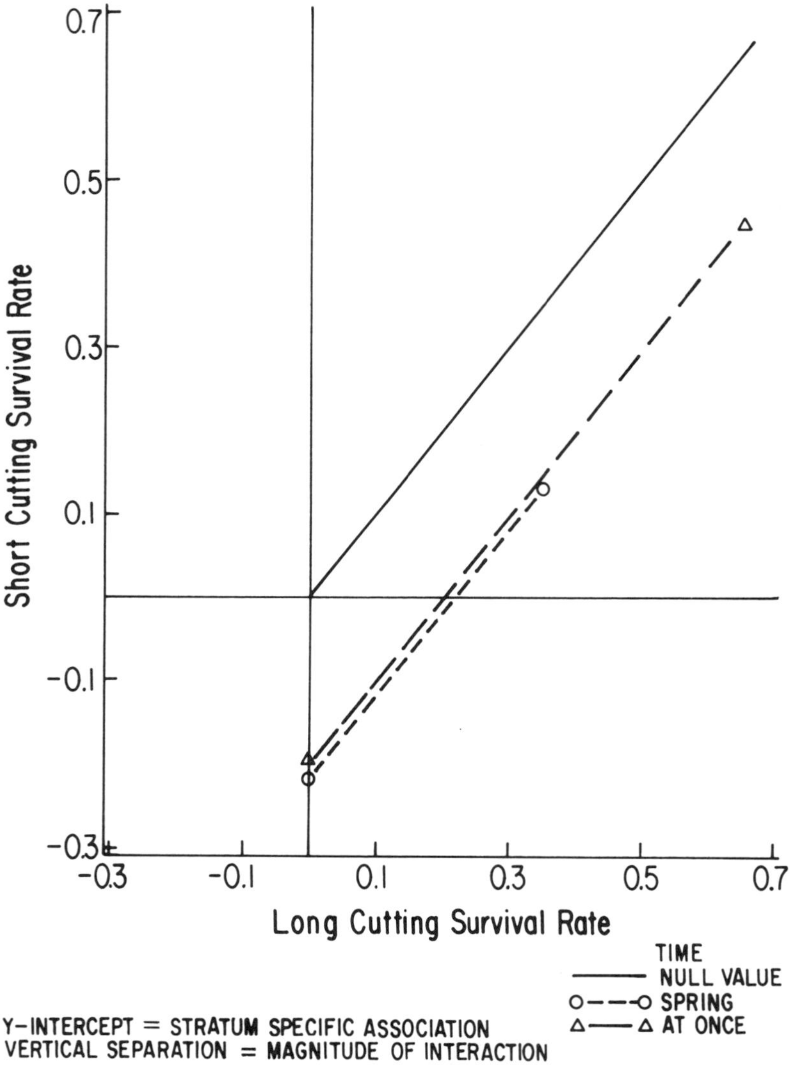

Figure 4.2 Plots of short vs. long cutting by survival.

since the corresponding odds ratios are *not* collapsible with respect to lineage.

Finally, the collapsibility theorems have important ramifications for study design. Suppose that *R, S,* and *T* correspond to the response, stratification, and treatment variables, respectively. Implicitly, the *RT* association is the only one of interest. Call it the *study association*. The study is stratified on *S* because prior studies suggest that it may affect the study association. In many circumstances we are willing to assume that no three-way interaction is present. Theorems 4.1 and 4.2 indicate that when *RT* is either the relative risk or the simple difference, it is collapsible with respect to *S* whenever *S* and *T* are independent. Hence we need only consider *S* variables which are known to be related to the treatment. Moreover, for randomized studies such as clinical trials, we can force the independence of *S* and *T*. An example of this is in the design that led to the data reported by Bartlett. These observations also apply to cross-sectional studies with a well-defined factor-response structure, and prospective studies where independence occurs "naturally."

The situation for unstructured cross-sectional studies and retrospective or case-control studies is somewhat different. Here the index of association of choice is the odds ratio. To assess the collapsibility of *RT* with respect to *S*, the conditional probabilities, $P(S = i|T = j, R = j)$, must be known. It follows that prior knowledge about the marginal associations provide little guidance as to the choice of *S*. Hence investigators using these designs are obligated to collect far more information than investigators using designs that can employ either the relative risk or simple difference. This also emphasizes the vulnerability of case-control study designs to the effects of unmeasured or excluded variables.

4.7 INTERPRETING AND SUMMARIZING ANALYSIS

The statistical analysis of contingency tables helps an investigator understand relationships among categorical variables. The use of test statistics and estimators permits statistical inferences about hypothetical associations. Alternatively, the construction of models can suggest new hypotheses which can be tested with fresh data sets. These two complementary approaches lead to the final and possibly most important step of interpreting statistical findings in a straightforward and uncomplicated manner. For two-factor, one-response tables this is reasonably easy. The response is either constant (homogeneous) or variable (heterogeneous) over the levels of the factors. The only difficult step is reinterpreting the response function in clear language.

The two- and three-response data sets are more difficult because the language of independence and odds ratios is generally less familiar, although in principle the interpretation is similar to that of the two-factor, one-response analysis. On the other hand, since unstructured analyses are inherently more ad hoc than structured analyses, and since case-control studies are more sensitive to the presence of excluded variables (see Section 4.6), this difficulty of interpretation should increase the investigator's caution in reporting findings.

The analyst has three points to cover in summarizing an analysis:

1. Choice of model
2. Tests of significance and parameter estimates
3. Appropriate summary tables

For choice of model, a clear indication of the model selection process should be given. In the case of the two-factor, one-response analysis, an indication of the selection of factors should be provided, together with a rationale for the response function. If one factor is the study or risk factor, the other factor (or factors, if several factors have been considered) can be justified by noting its association with the study factor. Similarly, the response function can be selected on purely substantive grounds. Alternatively, the function may be selected because it provides a simple yet adequate fit of the data to the model, that is, a model where the factors affect the response function in an additive manner.

For example, with Bartlett's data, because of the experimental design, planting time and length of cutting are independent on the margin. Hence it is appropriate to analyze the data as two 2×2 tables. On the other hand, there may be some concern that the two factors acted "synergistically." Hence a simple additive model is hypothesized for the data, with both factors being considered simultaneously. This model fit quite well, so the analyst can proceed with the hypothesis testing and estimation without worrying about the "synergism."

The other point concerns the use of the additive or simple difference response function. For Bartlett's data, this was selected on the quite arbitrary grounds that it came to mind first. On the other hand, epidemiologists frequently find the relative risk more appropriate. This would only require taking the logarithm of the survival rates and fitting the same parametric model. However, it turns out that the data do not fit this model in the sense of having a statistically significant error chi-square. The usual recommendation is to search for an alternative response function, such as the simple difference, which does not require a two-factor interaction term.

The second point requires reporting the tests of significance and estimates. A common format for two-factor, one-response models is shown in Table 4.15. Here the part (a) shows the test statistics that demonstrate the overall significance of the model, the significance of the individual factors, and the fit of the data to the model. Part (b) shows the estimated parameters and standard errors. Here interpretation requires explaining the selected identifiability conditions or equivalently, the dummy variables utilized. From the definitions of M, P, and L [(4.8) and (4.9) or (4.10)–(4.14)], it follows that early planting and long cuttings increase survival rates by 15.5 and 10.7 percentage points over the mean of 39.3 percent. Finally, the fitted values provided an explicit characterization of these effects (Part C).

Table 4.14 Estimated Model Parameters for Yanomama $(SV, VL)^{a}$

		Lineage				
Variable	Levels	Sha (1)	Hor (2)	Shamatar (3)	Other (4)	Main Effects (5)
Sex	Male	0.13	0.32**	−0.47**	0.02	−0.05
	Female	−0.13	−0.32**	0.47**	−0.02	0.05
Village	Lower	0.05	−1.15**	0.73**	0.37	−0.58**
	Upper	−0.09	−0.27	−0.04	0.40*	0.01
	Monou	−0.14	0.47*	0.31	−0.64**	−0.16
	Patanowa	0.18	0.95***	−1.00***	−0.13	0.74***
Main effects		0.42***	−0.08	−0.67***	0.33**	2.18

[a] z = Estimate/standard error: * $2 < z \leq 3$; ** $3 < z \leq 4$; *** $z > 4$.

As can be seen, tables such as Table 4.15 effectively combine points 2 and 3 for summarizing the analysis. A similar presentation is given in Tables 4.4 and 4.5 for the relationship between age, smoking, and SES. Notice first that smoking is heterogeneous between the two age groups, so both factors should be reported. Next, in the absence of prior information, an additive model was hypothesized. For purposes of illustration, three alternative response functions were considered, and all three gave similar results. The data clearly fit the model, age is nonsignificant, and smoking clearly is related to SES. Based on the parameterization (4.23), it appears that "former smokers" have the highest SES and "never smokers" have the lowest, regardless of the choice of averaging function.

The most complex data set that we have considered is the three-response data set for the Yanomama. Since the data were collected cross-sectionally and the order of precedence of the variables is unknown, the odds ratio is the preferred index of association. Moreover, the odds of selecting a male is the "sex ratio," a well-known demographic measure. Thus the log-linear model is appropriate. The model selection was based on backwards elimination of interaction terms in a hierarchical fashion, after computing all possible models. This is somewhat jargon filled, so the investigator is obligated to summarize the steps shown in Table 4.12. Here models 1–4 all "fit" the data in the sense of nonsignificant residual chi-square statistics. The rankit analysis suggests that model 2 overfits the data, so the *SV* term must be interpreted with caution (Table 4.9).

Next the estimated model is discussed. A common practice is to report the model as an equation, for example (4.58), and then tabulate the corresponding parameters as in Table 4.14. Here some ingenuity may be required to effectively display the possibly large number of parameters. The columns show first the interaction effects of lineage with sex and village [(1)–(4)], and finally the main effects for sex and village [column (5)]. The bottom row shows the main effect for lineage and the last entry is the overall mean ($\hat{M} = 2.18$). Many computer programs generate the large-sample standard errors, which permits a z-test or Wald test to be conducted. Thus we conclude that there is an excess of males among the Hor ($(SL)_{12} = 0.32$) and a deficit among the Shamatar ($(SL)_{13} = -0.47$) lineage. In Lower Bisaasi-teri there is an excess of other lineages ($(LV)_{42} = 0.40$); Monou and Patanowa show

Table 4.15 Weighted Least Squares Output for the Additive Model Applied to Bartlett's Data on Plum Root Stalks

a. Analysis of Variation

Source	df	Q	p-value
Model	2	207.11	0.0001
Time of planting	1	114.33	0.0001
Length of cutting	1	55.42	0.0001
Residual	1	0.08	0.7748
Total	3	207.19	0.0001

b. Estimated Parameters

Interpretation	Label	Estimate	Standard Error
Intercept	M	0.393	0.014
At once vs. spring	P	0.155	0.014
Long vs. short	L	0.107	0.014

c. Fitted Values

Group		Estimate	Standard Error
At once	Long	0.66	0.03
	Short	0.44	0.03
Spring	Long	0.34	0.03
	Short	0.13	0.02

an excess of Hor ($(LV)_{23} = 0.47$, $(LV)_{24} = 0.95$); Monou shows a deficit of other lineages ($(LV)_{43} = -0.64$. From this we conclude that there is a sharp difference in sex ratios between the Hor and Shamatar. Similarly, villages tend to show significant concentrations (and deficits) of lineages.

For many nonstatistical readers these observations may be difficult to follow. Here the collapsibility theorems suggest that a further aid may be provided by collapsing the sex by lineage table across village

(Table 4.7). Similarly, the village–lineage association may be validly displayed by collapsing the two sexes. Unfortunately, showing the independence of sex and village requires displaying the full three-dimensional table (Table 4.11).

APPENDIX 4A.1: WEIGHTED LEAST SQUARES ESTIMATORS AND TEST STATISTICS

Suppose that we have I multinomial distributions each with J levels. Let $\{p_{ij}\}$ be the set of multinomial probabilities corresponding to the ith distribution at level j and let n_i denote the ith sample size. Assume that $p_{ij} > 0$, $i = 1, \ldots, I$, $j = 1, \ldots, J$. Implicitly, $\Sigma_{j=1}^{J} p_{ij} = 1$. Let Y_{ij} be the corresponding multinomial random variable under the assumptions of Section 2.3. The expected value of Y_{ij} is

$$E(Y_{ij}) = n_i p_{ij} \qquad i = 1, \ldots, I, \quad j = 1, \ldots, J. \tag{4A.1}$$

The unbiased maximum likelihood estimator is

$$\hat{p}_{ij} = \frac{Y_{ij}}{n_i} \qquad i = 1, \ldots, I, \quad j = 1, \ldots, J. \tag{4A.2}$$

Let $\mathbf{p}_i' = (p_{i1}, p_{i2}, \ldots, p_{iJ})$ with corresponding vector estimator. The covariance matrix of p_i is

$$\mathbf{var}_i = \mathbf{var}\ (\mathbf{p}_i) = \frac{\mathbf{D}_{\mathbf{p}_i} - \mathbf{p}_i\mathbf{p}_i'}{n_i} \tag{4A.3}$$

where $\mathbf{D}_{\mathbf{p}_i}$ is a diagonal matrix with p_i on the diagonal. Let $\hat{\mathbf{p}}' = (\hat{p}_1, \hat{p}_2, \ldots, \hat{p}_I)$; the covariance matrix is given by

$$\mathbf{Var} = \begin{bmatrix} \mathbf{var}_1 & \mathbf{O} & \cdots & \mathbf{O} \\ \mathbf{O} & \mathbf{var}_2 & \cdots & \mathbf{O} \\ \cdots & \cdots & \cdots & \cdots \\ \mathbf{O} & \mathbf{O} & \cdots & \mathbf{var}_I \end{bmatrix}. \tag{4A.4}$$

The corresponding maximum likelihood estimators of $\mathbf{var}_i$ and $\mathbf{Var}$ may be obtained by replacing $\mathbf{p}$ with $\hat{\mathbf{p}}$ in (4A.3) and (4A.4). These are denoted $\mathbf{v}_i$ and $\mathbf{V}$, respectively.

Let $\mathbf{f}_h(\mathbf{p}_i) = \mathbf{f}_{hi}$ ($\neq 0$ for all $\mathbf{p}_i$) be a rational function of $\mathbf{p}_i$, where $h = 1, \ldots, H \leq I - 1$, and rational means $\mathbf{f}_{hi}$ is any function that can be constructed by an appropriate combination of addition (including addition of a constant), subtraction, multiplication, and division. Assume that $\mathbf{f}_{hi}$ are linearly independent, and let $\mathbf{f}(\mathbf{p}_i) = (\mathbf{f}_{1i}, \mathbf{f}_{2i}, \ldots, \mathbf{f}_{hI})'$. It can be shown that the large-sample covariance matrix of $\mathbf{f}(\mathbf{p}_i)$ is given by

$$\mathbf{V}(\mathbf{f}(\mathbf{p}_i)) = \mathbf{F}\ \mathbf{Var}\ (\mathbf{p}_i)\mathbf{F}', \tag{4A.5}$$

where $\mathbf{F} = [d\mathbf{f}_{hi}/d\mathbf{p}_i]|_{\mathbf{p}_i=\mathbf{p}_0}$. A proof is found in Bishop et al. (1975, Chap. 14).

Now suppose that the I distributions correspond to the combinations of several factors (independent variables). A model for the main effects and interactions may be written

$$E_A[\mathbf{f}(\mathbf{p})] = \mathbf{XB} \tag{4A.6}$$

where $E_A[\cdot]$ denotes large-sample expectation, $\mathbf{f}(\mathbf{p})$ is the $IH \times 1$ vector composed of the concatenated vectors $\mathbf{f}(\mathbf{p}_i)$, $\mathbf{B}$ is an unknown $(G \times 1)$ vector of parameters, and $\mathbf{X}$ is a matrix that characterizes the hypothesized study design (or the *design matrix*). Assume that $\mathbf{X}$ is of rank $G \leq IH$.

Let $\mathbf{W}$ be the covariance matrix estimator, which is a block diagonal matrix with $\mathbf{W}_i = \mathbf{V}(\mathbf{f}(\mathbf{p}_i))$ in the diagonal blocks and $H \times H$ blocks of zeros off the diagonal. The weighted least squares estimator of $\mathbf{B}$ is

$$\mathbf{b} = (\mathbf{X}'\mathbf{W}^{-1}\mathbf{X})^{-1}\mathbf{X}'\mathbf{W}^{-1}\mathbf{f}(\mathbf{p}). \tag{4A.7}$$

A Neyman chi-square test of the data fitting the model (4A.6) is based on

$$Q_e = (\mathbf{f}(\mathbf{p}) - \mathbf{Xb})'\mathbf{W}^{-1}(\mathbf{f}(\mathbf{p}) - \mathbf{Xb}) \tag{4A.8}$$

which will follow a chi-square distribution, with $IH - G$ degrees of freedom, when the model is appropriate. Note that if $G = IH$, the saturated model, then $Q_e = 0$.

It can also be shown that the large-sample covariance matrix estimator for $\mathbf{b}$ is given by

$$\mathbf{V(b)} = (\mathbf{X'W^{-1}X})^{-1}. \tag{4A.9}$$

Linear hypotheses about the **B** can be written

$$H_0\text{:}\quad \mathbf{CB} = \mathbf{O}. \tag{4A.10}$$

where **C** is of rank $c \leq G$. The Neyman chi-square is then

$$Q_c = (\mathbf{Cb})'(\mathbf{CV(b)C'})^{-1}\mathbf{Cb}. \tag{4A.11}$$

Given that the model (4A.6) is appropriate and the hypothesis (4A.10) is true, Q_c will also follow a chi-square distribution with df $= c$.

An important point is that the Neyman chi-square is the same as the well-known Wald (1943) test, based on several maximum likelihood estimates (Bhapkar 1966). To see this, let **L** be the orthogonal complement of **X**. It must be of rank $IH - G$. Then the hypothesis for the model (4A.6) is

$$H_0\text{:}\quad \mathbf{Lf(p)} = \mathbf{LXB} = \mathbf{O}. \tag{4A.12}$$

The corresponding covariance matrix is

$$\mathbf{V[Lf(p)]} = \mathbf{LWL'}. \tag{4A.13}$$

The Wald statistic is then

$$Q_W = [\mathbf{Lf(p)}]'(\mathbf{LWL'})^{-1}[\mathbf{Lf(p)}] \tag{4A.14}$$

which is chi-square df $= IH - G$ when H_0 (4A.12) is true. With a bit of matrix algebra it is easy to show that $Q_c = Q_W$. Notice that Q_c uses the weighted least squares estimate of **B**, while Q_W is based on the maximum likelihood estimates.

The formulation of appropriate functions is straightforward. For tests of homogeneity as with Bartlett's data, the index of association of choice is either the simple difference or relative risk. Let $p_i' = (p_{i1}, p_{i2})$. The rational functions are

$\mathbf{f(p}_i) = p_{1i} = (1 \quad 0)\mathbf{p}_i$ the simple difference and

$\mathbf{f(p}_i) = \mathbf{ln}\ (\mathbf{p}_i) = \mathbf{ln}\ [(1 \quad 0)(\mathbf{p}_i)]$ the relative risk.

The corresponding covariance matrix estimators are

$$\mathbf{V}(\mathbf{p}_i) = \mathbf{F}(\mathbf{p}_i)\mathbf{V}(\mathbf{p}_i)\mathbf{F}(\mathbf{p}_i)'$$
$$= (1 \quad 0)\begin{bmatrix} p_{i1}p_{i2} & -p_{i1}p_{i2} \\ -p_{i1}p_{i2} & p_{i1}p_{i2} \end{bmatrix}\begin{bmatrix} 1 \\ 0 \end{bmatrix}\frac{1}{n_i}$$
$$= \frac{p_{i1}p_{i2}}{n_i} \qquad \text{simple difference}$$

and

$$\mathbf{V}(\ln(\mathbf{p}_i)) = \frac{1}{p_{i1}}(1 \quad 0)\begin{bmatrix} p_{i1}p_{i2} & -p_{i1}p_{i2} \\ -p_{i1}p_{i2} & p_{i1}p_{i2} \end{bmatrix}\begin{bmatrix} 1 \\ 0 \end{bmatrix}\frac{1}{p_{i1}n_i}$$
$$= \frac{p_{i1}p_{i2}}{p_{i1}^2 n_i} = \frac{p_{i2}}{p_{i1}n_i} \qquad \text{relative risk.}$$

For both the corresponding design matrix is

$$X = \begin{bmatrix} 1 & 1 & 1 \\ 1 & 1 & -1 \\ 1 & -1 & 1 \\ 1 & -1 & -1 \end{bmatrix},$$

so the complete model for the simple difference model (4.5) is

$$E_A\begin{bmatrix} p_{11} \\ p_{21} \\ p_{31} \\ p_{41} \end{bmatrix} = \begin{bmatrix} 1 & 1 & 1 \\ 1 & 1 & -1 \\ 1 & -1 & 1 \\ 1 & -1 & -1 \end{bmatrix}\begin{bmatrix} M \\ P \\ L \end{bmatrix}.$$

A similar relationship holds for the relative risk.

In general, most rational functions of interest can be written in the form $\mathbf{f}(\mathbf{p}_i) = \mathbf{A}_3 \exp\{\mathbf{A}_2 \log[\mathbf{A}_1\mathbf{p}_i]\}$. Let $\mathbf{a}_{1i} = \mathbf{A}_1\mathbf{p}_i$ and $\mathbf{a}_{2i} = \exp\{\mathbf{A}_2 \log[\mathbf{A}_1\mathbf{p}_i]\}$. Correspondingly, $\mathbf{F}(\mathbf{p}_i) = \mathbf{A}_3\mathbf{D}_{\mathbf{a}_{2i}}\mathbf{A}_2\mathbf{D}_{\mathbf{a}_{1i}}\mathbf{A}_1$, so the large-sample covariance matrix is given by

$$\mathbf{Var}\,[\mathbf{f}(\mathbf{p}_i)] = \mathbf{A}_3\mathbf{D}_{\mathbf{a}_{2i}}\mathbf{A}_2\mathbf{D}_{\mathbf{a}_{1i}}\mathbf{A}_1(\mathbf{D}_{\mathbf{p}_i} - \mathbf{p}_i\mathbf{p}_i')\mathbf{A}_1'\mathbf{D}_{\mathbf{a}_{1i}}\mathbf{A}_2'\mathbf{D}_{\mathbf{a}_{2i}}\mathbf{A}_3'.$$

APPENDIX 4A.2: SAS CODE FOR ANALYZING BARTLETT'S DATA USING SIMPLE DIFFERENCE, RELATIVE RISK, AND LOG ODDS

```
TITLE 'ALTERNATIVE RESPONSE FUNCTIONS'
DATA;
  INPUT  C1-C8;                                 * READING THE 8 CELL COUNTS;
  ARRAY C  C1-C8;                               * CREATING VECTOR OF COUNTS;
  DO OVER C;
    COUNT = C;                                  * CREATING THE CELL COUNTS;
    SURVIVE = 2 - MOD(_I_,2);                   * SURVIVAL INDEX;
    LENGTH = 1+ FLOOR (MOD)(_I_-1,4)/2);* LENGTH OF CUTTING INDEX;
    PLANT = 1 + INT((I-1)/4);                   * TIME OF PLANTING INDEX;
    OUTPUT;
  END;
CARDS;
156 84 107 133 84 156 31 209
PROC FUNCAT;
  MODEL SURVIVE = LENGTH PLANT;                 * CONSTRUCTING DESIGN;
  RESPONSE 1 0;                                 * SIMPLE DIFFERENCE;
  RESPONSE 1 0 LOG;                             * RELATIVE RISK;
  RESPONSE 1 -1 LOG;                            * LOG ODDS;
  WEIGHT COUNT;
  CONTRAST 'MODEL' LENGTH 1, PLANT 1;           * TESTING TWO MODEL PARAMETERS;
```

APPENDIX 4A.3: ITERATIVE PROPORTIONAL FITTING

For three-way tables maximum likelihood estimates for models of the form

$$\ln(m_{ijk}) = M + A_i + B_j + C_k + AB_{ij} + AC_{ik} + BC_{jk} \quad \textit{usual identifiability conditions} \qquad (4A.15)$$

cannot be obtained directly. Iterative methods can be used. One of these, known as *iterative proportional fitting* or IPF, is due to Deming and Stephan (1943). Its statistical properties are found in Bishop et al. (1975). The important point is that when the data fits the model, the algorithm converges to the m.l.e.'s, and if direct estimates exist, convergence is in a single cycle (except for rounding error). Also, the algorithm establishes an equivalence between the model parameters and the marginal tables. Thus if *AB* is in the model, the *AB* margin is fit exactly.

Suppose that y_{ijk} are the observed values and $m_{ijk}^{r,s}$ are the estimates at the *r*th step of the *s*th cycle. To begin the algorithm set,

$$m_{ijk}^{0,0} = 1 \qquad i = 1, \ldots, I, \quad j = 1, \ldots, J, \quad k = 1, \ldots, K.$$

At step 1 we fit the *AB* margin:

$$m_{ijk}^{1,1} = \frac{m_{ijk}^{0,0} y_{ij+}}{m_{ij+}^{0,0}} = \frac{y_{ij+}}{k}.$$

Next we fit the *AC* margin:

$$m_{ijk}^{2,1} = \frac{m_{ijk}^{1,1} y_{i+k}}{m_{i+k}^{1,1}} = \frac{y_{ij+} y_{i+k}}{Y_{i++}}.$$

The final step in cycle 1 fits the *BC* margin:

$$m_{ijk}^{3,1} = \frac{m_{ijk}^{2,1} y_{+jk}}{m_{+jk}^{2,1}} = \frac{y_{+jk} y_{ij+} y_{ijk}}{(\Sigma_i y_{ij+} y_{i+k}/y_{i++}) y_{i++}}.$$

Table 4A.1 Steps and Cycles of IPF Algorithm for Infant Survival According To Duration of Prenatal Care Received by Mother (Table 4.8)

				m_{ijk}								
C	D	S	y_{ijk}	(0,0)	(1,1)	(2,1)	(3,1)	(1,2)	(2,2)	(3,2)	...	(4,3)
1	1	1	3	1	89.5	2.6323	2.6799	2.6817	2.7589	2.7735	...	2.8
1	1	2	176	1	89.5	176.37	176.20	176.32	176.23	176.19	...	176.2
1	2	1	4	1	148.5	4.3676	4.1236	4.1225	4.2411	4.1682	...	4.2
1	2	2	293	1	148.5	292.63	292.96	292.88	292.76	292.85	...	292.8
2	1	1	17	1	107	17.013	17.320	17.310	17.136	17.226	...	17.2
2	1	2	197	1	107	196.99	196.80	196.69	196.86	196.81	...	196.8
2	2	1	2	1	12.5	1.9874	1.8764	1.8828	1.8639	1.8318	...	1.8
2	2	2	23	1	12.5	23.013	23.039	23.118	23.138	23.145	...	23.2
1	1	+	179	2			178.88			178.96		
1	2	+	297	2			297.08			297.02		
2	1	+	214	2			214.12			214.04		
2	2	+	25	2			24.915			24.977		
1	+	1	7		238			6.8042				
1	+	2	469		238			469.20				
2	+	1	19		119.5			19.193				
2	+	2	220		119.5			219.81				
+	1	1	20			19.645			19.895			
+	1	2	373			373.36			373.09			
+	2	1	6			6.3550			6.1050			
+	2	2	316			315.64			315.90			

In general, at the sth cycle,

$$m_{ijk}^{1,s} = \frac{m_{ijk}^{3,s-1} \; y_{ij+}}{m_{ij+}^{3,s-1}}$$

$$m_{ijk}^{2,s} = \frac{m_{ijk}^{1,s} y_{i+k}}{m_{i+k}^{1,s}}$$

$$m_{ijk}^{3,s} = \frac{m_{ijk}^{2,s} y_{+jk}}{m_{+jk}^{2,s}}.$$

We now cycle through the steps until convergence occurs. As an example, consider the data in Table 4.8:

$$m_{111}^{1,1} = \frac{(1)179}{2} = 89.5 \qquad m_{112}^{1,1} = \frac{(1)179}{2} = 89.5$$

$$m_{121}^{1,1} = m_{122}^{1,1} = \frac{1(297)}{2} = 148.5 \qquad m_{211}^{1,1} = m_{212}^{1,1} = \frac{1(214)}{2} = 107$$

$$m_{221}^{1,1} = m_{222}^{1,1} = \frac{1(25)}{2} = 12.5.$$

These are shown as the first eight entries in the third column of Table 4A.1. Next the $m_{i+j}^{1,1}$ are computed. For example, $m_{1+1}^{1,1} = 238$.

In the second step of cycle 1 we obtain m$ijk@: for example, $m_{111}^{2,1} = 89.5(7)/238 = 2.6323$. These are shown in the fourth column together with the m_{+jk}. The stepping and cycling continues until the fitted values stabilize to the values shown in the last column after four cycles.

5

Multiple-Factor Models: GSK Methods

Chapter 4 covers the analysis of three categorical variables. Analyses begin by determining a factor-response structure. Next, an appropriate measure of association is selected and the analysis is undertaken. When the process involves only three variables, it is reasonably straightforward. When more than three variables are considered the process is conceptually the same, but operationally the details are more cumbersome.

As with three variables, the analysis begins by identifying the appropriate factor-response structure. Both the underlying study design and the substantive hypotheses of interest guide the identification. For example, in the study of the effect of length of cutting and time of planting on plum root stalk survival, the stalks were randomized into four treatment groups. It follows that time of planting and length of cutting should be treated as factors. The remaining question concerns the choice of response function. In this case, a simple one, the probability of survival given the treatment is selected.

In the example of infant survival, clinic, and duration of prenatal care, all variables were treated as responses because the underlying study design was unknown. This led to an analysis based on the log-linear model and the associated odds ratios. As will be seen in Chapter 7, logistic regression is also based on the odds-ratio measure of association, so the analysis is equivalent to treating clinic and duration of care as factors and using the ''logit'' response function.

For the remaining studies considered in Chapter 4, the underlying structure was less obvious. In the case of smoking, age, and SES, the first two were designated as factors, but it was recognized that this was somewhat artificial. Similarly, in the Yanomama example, all variables were treated as responses because the data were viewed as cross-sectional. Again it could be argued that village and lineage are fixed and therefore factors. The choice of the log odds as the response function in these cases meant that there would be no conflict between the alternative factor-response formulations.

This chapter focuses on data sets where two or more variables have been designated as factors. Attention is on fitting simple models for the association between factors and the response function. A major concern is a strategy of analysis when one or more of the factor's effect on the response depends on other factors. This is termed *modeling interaction.* The strategy depends heavily on the parameterization employed and therefore has a somewhat arbitrary quality.

It is recommended that the analyst divide the analysis into two phases. First, the basic study hypotheses should be tested. As a standard part of the test, assumptions should be evaluated. If any unanticipated interactions are discovered during this evaluation, it is important to report them as discovered rather than hypothesized, so as to preserve the probabilistic foundation of the basic significance tests. Of course, if certain interactions have been specified a priori, the selected parameterization will reflect this and the results interpreted accordingly.

In the second analysis phase, the model is reparameterized to provide a clearer and more parsimonious characterization of the interactions. At this point it is critical that the analyst have a clear understanding of the method of constructing various parameterizations. Moreover, it is equally important to recognize the arbitrariness of this reparameterization process. This is what is commonly referred to as *fitting the model to the data* or, less charitably, as ''data dredging.'' This recognition should induce extreme caution in the analyst's interpretation of the results of this phase. Ultimately, the findings of this phase require verification in subsequent studies.

All estimation and testing in this chapter are based on weighted least squares (WLS) and the Neyman chi-square or Wald test statistics. This procedure was first proposed by Grizzle, Starmer, and Koch (1969), hence the name GSK. The computations are performed using the SAS procedure FUNCAT* (SAS 1982b). Alternatively, maximum likelihood estimates and likelihood ratio test statistics may be obtained using the GLIM package (Baker and Nelder 1978). When the sample sizes are sufficiently large to justify any of the procedures, they all yield nearly identical results. When the results differ, the assumption of a large enough sample is probably inappropriate.

As noted in Chapter 2, a limitation of the WLS approach occurs in the presence of observed zeros ($y = 0$). It is common practice to replace these by either $1/r$, r = number of responses, or $1/rn$, n = sample size for the subpopulation. In general, the former is employed. As noted by a number of authors, this practice tends to make it more difficult to detect significant effects. On the other hand, it is less extreme than adding 0.5 to all cells as is recommended for improving the convergence of the m.l.e. iterative algorithms (Dixon 1981).

5.1 BASIC MODELS FOR THREE-FACTOR STRUCTURES

The data in Table 5.1 show knowledge of cancer cross-classified by subject's source of information. The cross-classification factors include newspaper readings, listening to the radio, solid reading, and attending lectures. The data set is quite old (Lombard and Doering 1947), so watching television was not relevant. These data have been extensively analyzed by many authors, including Winsor (1948), Dyke and Patterson (1952), Yates (1960, Sec. 9.7), Cox and Snell (1968), Bishop (1969), and Goodman (1978, pp. 111–141). The analysis will test the significance of each factor's association with cancer knowledge and then estimate the direction and magnitude of the association.

Since the proportion of good knowledge ranges from 0.167 to 0.742, analysis of simple proportions, log proportions, and log odds

*FUNCAT has been replaced by CATMOD in SAS version 5.

Table 5.1 Counts and Percentages for Good and Poor Cancer Knowledge by Newspaper Readership, Radio Listening, Reading, and Attending Lectures

			Attending Lectures and Cancer Knowledge					
			Yes			No		
Read Newspaper	Listen to Radio	Solid Reading	Good	Poor	Total	Good	Poor	Total
Yes	Yes	Yes	23	8	31	102	67	169
		%	74.2	25.8	100.0	60.4	39.6	100.0
		No	8	4	12	35	59	94
		%	66.7	33.3	100.0	37.2	62.8	100.0
	No	Yes	27	18	45	201	177	378
		%	60.0	40.0	100.0	53.2	46.8	100.0
		No	7	6	13	75	156	231
		%	53.8	46.2	100.0	32.5	67.5	100.0
No	Yes	Yes	1	3	4	16	16	32
		%	25.0	75.0	100.0	50.0	50.0	100.0
		No	4	3	7	13	50	63
		%	57.1	42.9	100.0	20.6	79.4	100.0
	No	Yes	3	8	11	67	83	150
		%	27.3	72.7	100.0	44.7	55.3	100.0
		No	2	10	12	84	393	477
		%	16.7	83.3	100.0	17.6	82.4	100.0

Source: Lombard and Doering (1947). Reprinted with permission from The Biometric Society.

will give approximately the same results. Arbitrarily, we analyze the simple proportion with good knowledge: $P(\text{good}|N = h, R = i, S = j, L = k) = q_{hijk} > 0$, where N, R, S, L correspond to newspaper readings, radio listening, solid reading, and attending lectures, respectively; and $h, i, j, k, = 1$ for yes and 2 for no.

The assumptions are:

1. The sample sizes, n_{hijk}, are treated as fixed.
2. The observations are independently obtained both within and across samples.
3. Within each sample the probability of good knowledge is constant (identical distribution).

These three assumptions mean that the data can be treated as being sampled from $2 \times 2 \times 2 \times 2 = 16$ binomial populations with associated random counts, Y_{hijk}. A fourth assumption, large sample size, is required for the validity of the statistical tests. For samples without lecture attendance ($L = 2$), this is clearly appropriate ($n_{hij2} > 30$), but for the remaining eight samples, especially those where subjects attend lectures but do not read newspapers, this is much more tenuous ($n_{2ij1} < 15$). Hence we need to be especially cautious about conclusions involving interactions between lecture attendance and the remaining variables.

The first phase of the analysis is to determine the significance and direction of the four primary associations. One possible model is

$$q_{hijk} = \mu + \nu_h + \rho_i + \sigma_j + \lambda_k \tag{5.1}$$

where Greek letters emphasize the nonidentifiable nature of the parameters. We can obtain an identifiable set by specifying five dummy variables:

$X_1 = 1$ always (coefficient for mean)

$X_2 = 1$ if $h = 1$, -1 if $h = 2$ (coefficient for newspaper reading)

$X_3 = 1$ if $i = 1$, -1 if $i = 2$ (coefficient for radio)

$X_4 = 1$ if $j = 1$, -1 if $j = 2$ (coefficient for solid reading)

$X_5 = 1$ if $k = 1$, -1 if $k = 2$ (coefficient for lecture attendance).

The reparameterized model is then

$$q_{hijk} = X_1M + X_2N + X_3R + X_4S + X_5L. \tag{5.2}$$

Here M is the overall mean, while N, R, S, and L are deviations from this mean due to newspapers, radio, solid reading, and lectures, respectively. The detailed set of variable values and coefficients is shown in the first nine columns of Table 5.2.

The set of coefficient values shown in Table 5.2 completely specify the 16 samples or the experimental design. For this reason such a matrix is often termed a *design matrix*. Sometimes it is more convenient to use matrix notation. Let $q' = (q_{1111}, q_{1112}, q_{1121}, \ldots,$

Table 5.2 Design Matrix for Main-Effects and Saturated Models

Variable				Coefficients																
N (1)	R (2)	S (3)	L (4)	X_1 (5)	X_2 (6)	X_3 (7)	X_4 (8)	X_5 (9)	X_6 (10)	X_7 (11)	X_8 (12)	X_9 (13)	X_{10} (14)	X_{11} (15)	X_{12} (16)	X_{13} (17)	X_{14} (18)	X_{15} (19)	X_{16} (20)	**B**
1	1	1	1	1	1	1	1	1	1	1	1	1	1	1	1	1	1	1	1	*M*
1	1	1	2	1	1	1	1	−1	1	1	−1	1	−1	−1	1	−1	−1	−1	−1	*N*
1	1	2	1	1	1	1	−1	1	1	−1	1	−1	1	−1	−1	1	−1	−1	−1	*R*
1	1	2	2	1	1	1	−1	−1	1	−1	−1	−1	−1	1	−1	−1	1	1	1	*S*
1	2	1	1	1	1	−1	1	1	−1	−1	1	−1	−1	1	−1	−1	1	−1	−1	*L*
1	2	1	2	1	1	−1	1	−1	−1	−1	−1	−1	1	−1	−1	1	−1	1	1	*NR*
1	2	2	1	1	1	−1	−1	1	−1	1	1	1	−1	−1	1	−1	−1	1	1	*NS*
1	2	2	2	1	1	−1	−1	−1	−1	1	−1	1	1	1	1	1	1	−1	−1	*NL*
2	1	1	1	1	−1	1	1	1	−1	1	−1	1	1	1	−1	−1	−1	1	−1	*RS*
2	1	1	2	1	−1	1	1	−1	−1	1	1	1	−1	−1	−1	−1	1	−1	1	*RL*
2	1	2	1	1	−1	1	−1	1	−1	−1	−1	−1	1	−1	1	−1	1	−1	1	*SL*
2	1	2	2	1	−1	1	−1	−1	−1	−1	1	−1	−1	1	1	1	−1	1	−1	*NRS*
2	2	1	1	1	−1	−1	1	1	1	−1	−1	−1	−1	1	1	1	−1	−1	1	*NRL*
2	2	1	2	1	−1	−1	1	−1	1	−1	1	−1	1	−1	1	−1	1	1	−1	*NSL*
2	2	2	1	1	−1	−1	−1	1	1	1	−1	1	−1	−1	−1	1	1	1	−1	*RSL*
2	2	2	2	1	−1	−1	−1	−1	1	1	1	1	1	1	−1	−1	−1	−1	−1	*NRSL*
				Main Effects					Pairwise (Interactions)						Triple (Interactions)				Four-Way (Interactions)	

q_{2222}) and $B' = (m, N, R, S, L)$. The matrix representation of the model is then

$$\mathbf{q} = \mathbf{X}_M \mathbf{B}, \tag{5.3}$$

where $\mathbf{X}_M$ is the 16×5 matrix of coefficients in columns (5)–(9) of Table 5.2. The subscript M indicates that this is the main-effects model.

The estimation of **B** and the corresponding test statistics are based on the usual weighted least squares computations discussed in Appendix 4A.1. The results are tabulated in columns (2) and (3) of Table 5.3. Here "-" indicates either that a term is included or a test is inappropriate. The overall fit of the data to the model is in the row labeled "residual." There $Q = 12.72$, df $= 11$, which corresponds to $p > 0.25$, so these data are consistent with the main-effect model.

The parameter estimates and test statistics are shown in the first five rows of columns (2) and (3). All of the Q statistics exceed the 95th percentile of the chi-square distribution with df $= 1$, so all four effects are declared to be statistically significant. Since the parameter coefficients are 1 when the variable values are "yes," doing any of the activities is associated with a larger proportion of subjects with good knowledge. Finally, solid reading is both largest in magnitude and has the smallest p-value. Attending lectures has the largest p-value but is larger in magnitude than listening to radio. This is probably due to the small sample sizes for attending lectures.

This completes phase 1 of the analysis except for evaluating the underlying assumptions of the procedures. This can be done, in part, with residual analysis, as discussed in Chapter 4. Since there are 16 samples and 5 parameters, only 11 residuals should be considered. A rankit plot of the standardized residuals reveals neither major outliers nor any clear patterns of deviations from normality. But the residual chi-square is large enough to indicate the inclusion of one or more significant parameters. With this in mind, phase 2 of the analysis, complex modeling, or fitting of a model to the data, is undertaken.

A first step is fitting a saturated model, one with 16 parameters, to the data. The coefficients of the interaction terms may be constructed as shown in Table 5.4. The detailed values of the coefficients are in columns (10)–(20) of Table 5.2. Common names for these coefficients are given at the bottom of Table 5.2.

Table 5.3 Analysis of Variation and Estimates for Data in Table 5.1, Four Factors–One Response (Good Knowledge of Cancer)[a]

Parameter Label (1)	Main Effects Only: Estimate (2)	Main Effects Only: Q (3)	Saturated Model: Estimate (4)	Saturated Model: Q (5)
M	0.436	—	0.436	—
N	0.072	34.37	0.112	17.94
R	0.033	6.08	0.053	4.10
S	0.111	79.24	0.058	4.78
L	0.043	4.34	0.040	2.34
NR	—	—	−0.005	0.03
NS	—	—	0.014	0.29
NL	—	—	0.049	3.44
RS	—	—	−0.023	0.75
RL	—	—	0.028	1.13
SL	—	—	−0.068	6.55
NRS	—	—	0.028	1.10
NRL	—	—	−0.009	0.12
NSL	—	—	0.030	1.28
RSL	—	—	−0.029	1.19
$NRSL$	—	—	0.028	1.09
Model	df = 4	248.32	df = 15	261.04
Residual	df = 11	12.72	df = 0	—
Total	df = 15	261.04	df = 15	261.04

[a]—, Deleted term or inappropriate test.

Since the main effects correspond to the associations between each factor and the response after adjusting for the other factors, these are termed *partial associations*. The next six terms correspond to the association between one factor and the response depending on a second factor. These are *pairwise interactions of factors*. For the multiple response or independence models the corresponding parameters are termed *three-factor interactions*. Similarly, the next four terms [col-

Table 5.4 Coefficients of Interaction Terms Based on Main-Effect Identifiability Conditions for Model (5.2)

Interaction	Coefficient	Interaction	Coefficient
NR	$X_6 = X_2X_3$	*NRS*	$X_{12} = X_2X_3X_4$
NS	$X_7 = X_2X_4$	*NRL*	$X_{13} = X_2X_3X_5$
NL	$X_8 = X_2X_5$	*NSL*	$X_{14} = X_2X_4X_5$
RS	$X_9 = X_3X_4$	*RSL*	$X_{15} = X_3X_4X_5$
RL	$X_{10} = X_3X_5$	*NRSL*	$X_{16} = X_2X_3X_4X_5$
SL	$X_{11} = X_4X_5$		

umns (15)–(19)] represent a dependence of the factor-response association on two other factors. Finally, the 16th coefficient [column (20)] corresponds to the *four-factor interaction.*

The estimates and corresponding tests are shown in Table 5.3 for the saturated model. Since there are 16 samples and 16 estimates there is no residual variation. The model test is therefore equal to the total variation.

As before, the individual terms can be tested by computing the Wald statistics:

$$Q = (z)^2 = \frac{(b_d)^2}{V(b_d)} \qquad \text{df} = 1, \tag{5.4}$$

where b_d is the estimate of the dth parameter, $d = 1, \ldots, 16$. Here the estimators of B_d and Var (b_d) are contained in Appendix 4A.1. These tests must be interpreted carefully since so many nonsignificant parameters have been included in the model, which tends to reduce both the power of the test and the immediate interpretability of the results.

The first point of interest is that *N*, *R*, and *S* are still statistically significant but for *L*, $p > 0.10$. Also, the p-value for both *N* and *S* has increased dramatically. The second point is that all of the three- and four-factor interactions are quite small ($p > 0.25$). However, the p-values for *NL* and *SL* do approach the usual significance levels ($p < 0.10$, $p < 0.025$), respectively. This suggests that radio listening acts only as a main effect, but both newspaper reading and solid reading are modified by attending lectures.

This suggestion can be formalized by conducting *joint hypothesis tests* of the parameters. By "joint hypothesis" is meant the simultaneous testing of several one-degree-of-freedom hypotheses. Specifically, hypotheses of the form H_0: $B_1 = B_2 = \cdots = B_c = 0$ are of interest. For example, the joint hypothesis of no three- or four-factor interaction is

$$H_{3-4}\colon \quad NRS = NRL = NSL = RSL = NRSL = 0. \tag{5.5}$$

Since five parameters are set equal to zero (implicitly), only 11 parameters need to be estimated: df = 16 − 11 = 5. Similarly, the test of no two-factor or higher interaction is H_{2-3-4}: $NR = NS = \cdots = RSL = NRSL = 0$. Now 11 parameters are set equal to zero, so df = 11.

The discussion of joint tests is simplified by rewriting these hypotheses in matrix terms. For example, the hypothesis of no three- or four-factor interaction is written as

$$H_{3-4}\colon \begin{bmatrix} 0&0&0&0&0&0&0&0&0&0&0&1&0&0&0&0 \\ 0&0&0&0&0&0&0&0&0&0&0&0&1&0&0&0 \\ 0&0&0&0&0&0&0&0&0&0&0&0&0&1&0&0 \\ 0&0&0&0&0&0&0&0&0&0&0&0&0&0&1&0 \\ 0&0&0&0&0&0&0&0&0&0&0&0&0&0&0&1 \end{bmatrix} \begin{bmatrix} M \\ N \\ R \\ S \\ L \\ NR \\ NS \\ NL \\ RS \\ RL \\ SL \\ NRS \\ NRL \\ NSL \\ RSL \\ NRSL \end{bmatrix} = \mathbf{CB} = 0. \tag{5.6}$$

The **C** matrix is often termed a *contrast matrix*. Notice that it has 16 columns corresponding to the 16 rows of **B**. Further, it has one row for each simple hypothesis contained in the joint hypothesis. This row multiplied by the **B** vector then yields the corresponding simple hypothesis. Any number of simple hypotheses can be tested simultaneously as long as none is implied by two or more of the remainder in general. More concretely, we require that **C**, a $c \times d$ matrix, be of rank c, with d being the number of parameters in the model. As another example, consider the contrast matrix for H_{2-3-4}:

$$\mathbf{C} = \begin{bmatrix} 0 & 0 & 0 & 0 & 0 & 1 & 0 & 0 & 0 & 0 & 0 & 0 & 0 & 0 & 0 & 0 \\ 0 & 0 & 0 & 0 & 0 & 0 & 1 & 0 & 0 & 0 & 0 & 0 & 0 & 0 & 0 & 0 \\ 0 & 0 & 0 & 0 & 0 & 0 & 0 & 1 & 0 & 0 & 0 & 0 & 0 & 0 & 0 & 0 \\ 0 & 0 & 0 & 0 & 0 & 0 & 0 & 0 & 1 & 0 & 0 & 0 & 0 & 0 & 0 & 0 \\ \cdots & & & & & & & \cdots & & & & & & & \cdots & \\ 0 & 0 & 0 & 0 & 0 & 0 & 0 & 0 & 0 & 0 & 0 & 0 & 0 & 0 & 1 & 0 \\ 0 & 0 & 0 & 0 & 0 & 0 & 0 & 0 & 0 & 0 & 0 & 0 & 0 & 0 & 0 & 1 \end{bmatrix}$$

This has 11 rows and no two add to the third, so the rank is 11, corresponding to the degrees of freedom. Other contrast matrices corresponding to complex relationships among the parameters can be constructed similarly.

A variety of joint hypothesis tests are shown in Table 5.5. These actually begin with the no-four-factor-interaction simple hypothesis, which can be accepted. Similarly, the joint hypothesis of no three- or four-factor interaction can be accepted. Next the hypothesis of no two-, three-, or four-factor interaction is identical to the residual for the main effect model. This hypothesis can also be accepted. However, notice that the test statistic is much larger than for the previous hypotheses. It is possible that it includes tests of one or more nonzero parameters. This possibility is examined by considering the next four joint hypotheses.

First, notice that the no-radio-interaction hypothesis test statistic is nonsignificant even on one degree of freedom. The remaining joint hypothesis statistics are large enough to contain one or more significant test statistics. From this we again conclude that it is possible that one or both of *NL* and *SL* should be included in a model. Moreover, it is evident that radio listening acts only as a main effect. This is basically the same set of conclusions as those reached by analyzing the saturated model.

Table 5.5 Joint Tests Based on Saturated Model for Cancer Knowledge Data

Hypothesis	df	Q	Interpretation
$NRSL = 0$	1	1.09	No four-factor interaction
$NRS = NRL = NSL = RSL = NRSL = 0$	5	2.58	No three- or four-factor interaction
$NR = NS = NL = RS = RL = SL = 0$	11	12.72	No two-, three-, or four-factor interaction
$NRS = NRL = NSL = RSL = NRSC = 0$			
$NR = NS = NL = NRS = NRL = NSL = NRSL$	7	7.25	No newspaper interaction
$NR = RS = RL = NRS = NRL = RSL = NRSL$	7	3.76	No radio interaction
$NS = RS = SL = NRS = NSL = RSL = NRSL$	7	7.85	No reading interaction
$NL = RL = SL = NRL = NSL = RSL = NRSL$	7	11.63	No lecture interaction
$N = NR = NS = NL = NRS = NRL = NSL = NRSL$	8	9.48	Total radio effects
$S = NS = RS = SL = NRS = NSL = RSL = NRSL$	8	88.18	Total reading effects
$L = NL = RL = SL = NRL = NSC = RSL = NRSL$	8	15.94	Total lecture effects

These two analyses suggest hypothesizing a model for subsequent study which contains main effects for all variables as well as the *NL* and *SL* two-variable interactions. The formal testing of such a model requires new or independently collected data. The formation of an interpretable parameterization for these two interactions is discussed in the next section.

5.2 MODULAR MODELS FOR INTERACTION

The hypothesis of an interaction between two or more factors is less commonly encountered in study design than hypotheses of heterogeneity and association. This is because in studying particular phenomena, attention tends to focus on the particular associations as well as those variables that cannot be safely collapsed. Moreover, the interaction may depend on the particular response function or measure of association considered. For example, an interaction found between two factors for prevalence may be removed by a log or log-odds transformation. This phenomenon has been of considerable interest to statisticians analyzing continuous data (e.g., Draper and Smith 1981). The same possibility exists for categorical data (Walter and Holford 1978). During the phase 2 analysis of a data set (model fitting), such possibilities should always be explored before hypothesizing a particular form of the interaction.

Phenomena with a long history of study may well have hypothesized interactions. Byssinosis is one such phenomenon. It is a respiratory ailment confined almost exclusively to the textile industry, especially cotton mill workers. It has been claimed, among other things, that the disease is actually induced by smoking rather than working in a dusty environment. This implies the possibility of an interaction between smoking and work environment. Moreover, the disease takes many years to develop, so most studies have focused on the prevalence of the disease. Hence we have specified, a priori, both the response function of interest and the presence of one or more interaction effects on this response.

A cross-tabulation of byssinosis prevalence by place and years of employment and smoking is shown in Table 5.6. This is a simplified form of the data sets used for studying the disease. Let q_{ijk} denote the proportion of workers in dusty area i, years of employment j, and

Table 5.6 Byssinosis Prevalence (Counts and Percents) by Working in Dusty Areas, Years Employed, and Smoking Habits

		Byssinosis			
		Smokers		Nonsmokers	
Dusty Area	Years Worked	%	Sample	%	Sample
Yes	<10	5.6	126	12.9	233
	10–<20	15.0	20	23.9	67
	20+	11.1	72	27.2	151
No	<10	1.2	1016	1.0	1354
	10–<20	0.9	211	1.2	414
	20+	1.0	785	2.0	970

Source: Higgins and Koch (1977).

smoking status k, where D, Y, and S denote the respective factors. The reparameterized model with the hypothesized two-factor interactions is

$$q = M + X_2D + X_3Y_1 + X_4Y_2 + X_5S + X_6(DY_1) + X_7(DY_2) + X_8(DS) + X_9(Y_1S) + X_{10}(Y_2S), \qquad (5.7)$$

where

$$X_2 = 1 \text{ if } i = 1, \quad -1 \text{ if } i = 2,$$

$$X_3 = 1 \text{ if } j = 1, \quad 0 \text{ if } j = 2, \quad -1 \text{ if } j = 3,$$

$$X_4 = 0 \text{ if } j = 1, \quad 1 \text{ if } j = 2, \quad -1 \text{ if } j = 3,$$

$$X_5 = 1 \text{ if } k = 1, \quad -1 \text{ if } k = 2,$$

$$X_6 = X_2X_3 \quad X_7 = X_2X_4 \quad X_8 = X_2X_5 \quad X_9 = X_3X_5 \quad X_{10} = X_4X_5.$$

Notice that the year-worked variable has three levels, so two dummy variables are required. The detailed set of coefficients is shown in Table 5.7.

Table 5.7 Design Matrices for Preliminary and Modular Models of Byssinosis Prevalence

Dusty	Years	Smoking	X_1	X_2	X_3	X_4	X_5	X_6	X_7	X_8	X_9	X_{10}	**B**
						Initial Model							
Y	<10	*N*	1	1	1	0	1	1	0	1	1	0	*M*
Y	<10	*Y*	1	1	1	0	−1	1	0	−1	−1	0	*D*
Y	10–<20	*N*	1	1	0	1	1	0	1	1	0	1	Y_1
Y	10–<20	*Y*	1	1	0	1	−1	0	1	−1	0	−1	Y_2
Y	20+	*N*	1	1	−1	−1	1	−1	−1	1	−1	−1	*S*
Y	20+	*Y*	1	1	−1	−1	−1	−1	−1	−1	1	1	DY_1
N	<10	*N*	1	−1	1	0	1	−1	0	−1	1	0	DY_2
N	<10	*Y*	1	−1	1	0	−1	−1	0	1	−1	0	*DS*
N	10–<20	*N*	1	−1	0	1	1	0	−1	−1	0	1	SY_1
N	10–<20	*Y*	1	−1	0	1	−1	0	−1	1	0	−1	SY_2
N	20+	*N*	1	−1	−1	−1	1	1	1	−1	−1	−1	
N	20+	*Y*	1	−1	−1	−1	−1	1	1	1	1	1	
					Modular Model: Y(D = *Yes*) S(D = *Yes*)								
Y	<10	*N*	1	1	1	0	1						*M*
Y		*Y*	1	1	0	1	1						*D*
Y	10–<20	*N*	1	1	0	1	1						$Y_1(D = 1)$
Y		*Y*	1	1	0	1	−1						$Y_2(D = 1)$
Y	20+	*N*	1	1	−1	−1	1						$S(D = 1)$
Y		*Y*	1	1	−1	−1	−1						
N	<10	*N*	1	−1	0	0	0						
N		*Y*	1	−1	0	0	0						
N	10–<20	*N*	1	−1	0	0	0						
N		*Y*	1	−1	0	0	0						
N	20+	*N*	1	1	0	0	0						
N		*Y*	1	−1	0	0	0						

The assumptions of independent samples of fixed size and identical distributions within samples leads to considering the data as a set of 12 binomials. The usual weighted least squares computations lead to the test statistics and estimates shown in Table 5.8. Since there are 10 estimated parameters and 12 samples, the residual has df = 12 − 10 = 2. $Q = 1.61$, so the data fit the model of no-three-factor interaction.

Given that the model is accepted, the remaining hypotheses may be considered. It is evident that both smoking and years employed interacts with working in a dusty area. However, smoking and years employed do not interact. This suggests that different models are appropriate depending on work area. Without affecting the significance levels of the basic tests, such a model may be formulated as

$$q_{ijk} = M + X_2D + X_3[Y_1(D = 1)] + X_4[Y_2(D = 1)] + X_5[S(D = 1)] + X_6[Y_1(D = 2)] + X_7[Y_2(D = 2)] + X_8[S(D = 2)], \quad (5.8)$$

where

X_2 as before

$X_3 = 1$ if $i = 1$ and $j = 1$, -1 if $i = 1$ and $j = 3$, 0 otherwise

$X_4 = 1$ if $i = 1$ and $j = 2$, -1 if $i = 1$ and $j = 3$, 0 otherwise

$X_5 = 1$ if $i = 1$ and $k = 1$, -1 if $i = 1$ and $k = 2$, 0 otherwise

$X_6 = 1$ if $i = 2$ and $j = 1$, -1 if $i = 2$ and $j = 3$, 0 otherwise

$X_7 = 1$ if $i = 2$ and $j = 2$, -1 if $i = 2$ and $j = 3$, 0 otherwise

$X_8 = 1$ if $i = 2$ and $k = 1$, -1 if $i = 2$ and $k = 2$, 0 otherwise.

Here the nonsignificant *YS* terms have been deleted. Notice that $X_3 - X_5$ are zero unless $i = 1$, so they are the coefficients of the Years and Smoking effects in the dusty areas. In other words, they are the parameter coefficients in the "dusty area module." Similarly, $X_6 - X_8$ are zero unless $i = 2$, so they correspond to the "non-dusty area module."

The joint test of H_0: $Y1(D = 2) = Y2(D = 2) = S(D = 2) = 0$ yields $Q = 1.45$ (df = 3), so these terms are clearly nonsignificant as well. This leads to the model containing only the first five terms in (5.8). The detailed design matrix is shown in Table 5.7 under the label "Modular Model." The corresponding test statistics and estimates are

Table 5.8 Analysis of Variation for Initial Model of Byssinosis Prevalence (Table 5.6)

Source	df	Q	p-Value
Model	9	114.35	0.00
Dusty (D)	1	68.26	0.00
Years (Y)	2	14.20	0.00
Smoking (S)	1	15.37	0.00
$D \cdot Y$	2	12.71	0.00
$D \cdot S$	1	13.68	0.00
$Y \cdot S$	2	2.80	0.25
Residual	2	1.61	0.45
Total	11	115.96	0.00

in Table 5.9. The data fit this model ($Q = 5.80$, df $= 7$). The parameter estimates indicate:

1. Working in dusty areas increases the prevalence ($D = 0.074$).
2. Working <10 years in dusty areas leads to lower prevalence than longer periods [$Y1(D = 1) = -0.066$].
3. Not smoking decreases the prevalence.

It might be argued that the "modular model" was, in part, a phase 2 (model fitting) analysis, but it is important to recognize that both of the

Table 5.9 Analysis of Variation and Parameter Estimates for Modular Model of Byssinosis Prevalence (Table 5.6)

Source	df	Q	p-Value	Label	Parameter Estimate	S.E.
Model	4	110.10	0.00	M	0.086	0.009
D	1	69.01	0.00	D	0.074	0.009
$Y(D = 1)$	2	13.64	0.00	$Y_1(D = 1)$	−0.066	0.020
$S(D = 1)$	1	14.30	0.00	$Y_2(D = 1)$	0.034	0.031
				$S(D = 1)$	−0.047	0.013
Residual	7	5.80	0.56			
Total	11	115.90	0.00			

interactions were hypothesized a priori, whereas *YS* was not. Hence (5.8) was, in effect, hypothesized a priori. The only questionable maneuver was deleting $X_6 - X_8$ from the design. Whether they are included or not, the overall results are essentially unchanged.

5.3 INCOMPLETE FACTORIAL DESIGNS

In Sections 5.1 and 5.2 there has always been one sample for each combination of factors. That is, suppose that factors *A*, *B*, and *C* have levels *I*, *J*, and *K*; then there have been *IJK* samples of size $n_{ijk} > 0$ for $i = 1, \ldots, I, j = 1, \ldots, J$, and $k = 1, \ldots, K$. These are termed *complete factorial designs*. Analysts of observational studies are not always so fortunate.

An example of an *incomplete factorial design* is shown in Table 5.10. Here the number of pediatric physician contacts in seven ordinal categories is cross-classified by the subscriber's age (*A*), number of children (*N*), and age of the youngest child (*Y*). These three factors have three, two, and three levels, respectively, for which a complete factorial design would imply 18 samples. However, there are no subscribers <25 who have the youngest child age 5 or more. Similarly, in the oldest age group the youngest child is always 5 or more. Altogether there are only 9 samples rather than 18 samples for analysis.

We say the design is incomplete to emphasize that the missing cells are not inherently zero, but they were not observed in this sample of 866 subscribers. These *sampling zeros* are distinct from *structural zeros,* which occurred when cells were deleted from contingency tables while testing for quasi-independence. Structural zeros can occur naturally as well. Fortunately, for data with a single response function the analysis is straightforward. For multiple-response data sets with structural zeros special methods and hypotheses must be employed (Bishop et al. 1975).

Having identified the factors, the next step is to specify the response function. Since pediatric contacts is an ordinal response, the mean score based on the midpoints of the categories seems reasonable. Alternative scales were discussed in Sections 3.3.2 and 4.2. The scores for the category midpoints are $\mathbf{a}' = (a_1, a_2, a_3, a_4, a_5, a_6, a_7) = (0, 1, 2, 3, 5, 8, 20)$. Here the final score, $a_7 = 20$, is chosen arbitrarily to reflect that category being the mode of each of the first six

Table 5.10 Frequencies for Pediatric Physicians Contacts by Subscriber's Age, Number of Enrolled Children, and Age of Youngest Child

Age of Subscriber (1)	Number of Children (2)	Age of Youngest (3)	Number of Pediatric Contacts 0 (4)	1 (5)	2 (6)	3 (7)	4–6 (8)	7–9 (9)	10+ (10)	Total (11)
<25	1	<5	4	8	2	5	10	8	19	56
	2 +	<5	3	2	2	1	8	8	25	49
25–44	1	<5	3	2	4	7	14	7	26	63
	2 +	<5	17	7	9	18	37	41	172	301
	2 +	5–9	17	22	12	19	33	9	46	158
	2 +	10+	25	5	4	7	5	5	9	60
45–59	1	10+	63	5	5	5	4	3	3	88
	2 +	5–9	5	5	7	3	8	1	4	33
	2 +	10+	32	4	6	2	6	5	3	58

Source: Forthofer and Lehnen (1981, Table 6.6).

samples. Ideally, the actual mean visits in the highest category should be used.

This response function yields the grouped mean pediatric contacts for each of the nine samples. For example, in the sample where age <25, number >2, youngest <5, $f_{121} = \mathbf{a}'\mathbf{p}_{121} = [0(3) + 1(2) + 2(2) + 3(1) + 5(8) + 8(8) + 20(25)]/49 = 12.51$. The usual assumptions lead to assuming contacts follow a multinomial, and from this the standard error of f_{ijk} can be computed.

The next step in the analysis is to choose a model and the corresponding parameterization. Since we have no prior information, a main-effects model is appropriate for the initial phase of the analysis. This means a model of the form

$$f_{ijk} = m + X_2A_1 + X_3A_2 + X_4N + X_5Y_1 + X_6Y_2, \tag{5.9}$$

where

$$X_2 = 1 \text{ if } i = 1, \quad -1 \text{ if } i = 3, \quad 0 \text{ otherwise}$$

$$X_3 = 1 \text{ if } i = 2, \quad -1 \text{ if } i = 3, \quad 0 \text{ otherwise}$$

$$X_4 = 1 \text{ if } j = 1, \quad -1 \text{ if } j = 2,$$

$$X_5 = 1 \text{ if } k = 1, \quad -1 \text{ if } k = 3, \quad 0 \text{ otherwise}$$

$$X_6 = 1 \text{ if } k = 2, \quad -1 \text{ if } k = 3, \quad 0 \text{ otherwise.}$$

The detailed design is shown in Table 5.11, where the unobserved samples are indicated by a solid line.

There are 9 samples and 6 parameters in the main effects model, so this leaves $9 - 6 = 3$ degrees of freedom for testing the fit of the data to the model. The corresponding chi-square residual is shown in Table 5.12 ($Q = 3.43$), which is not only nonsignificant ($p = 0.33$) but is sufficiently small to indicate that the test of any single excluded parameter would also be nonsignificant.

Table 5.11 Main-Effect Design Matrix for Pediatric Contacts, Indicating Deleted Rows

Age (i)	Number (j)	Youngest (k)	X_1	X_2	X_3	X_4	X_5	X_6	**B**
<25	1	<5	1	1	0	1	1	0	M
		5–9	—	—	—	—	—	—	$A1$
		10+	—	—	—	—	—	—	$A2$
	2 +	<5	1	1	0	−1	1	0	N
		5–9	—	—	—	—	—	—	Y_1
		10+	—	—	—	—	—	—	Y_2
25–44	1	<5	1	0	1	1	1	0	
		5–9	—	—	—	—	—	—	
		10+	—	—	—	—	—	—	
	2 +	<5	1	0	1	−1	1	0	
		5–9	1	0	1	−1	0	1	
		10+	1	0	1	−1	−1	−1	
45–59	1	<5	—	—	—	—	—	—	
		5–9	—	—	—	—	—	—	
		10+	1	−1	−1	1	−1	−1	
	2 +	<5	—	—	—	—	—	—	
		5–9	1	−1	−1	−1	0	1	
		10+	1	−1	−1	−1	−1	−1	

Table 5.12 Test Statistics and Estimates for Pediatric Contact Data

Source	df	Q	p-Value	Parameter Label	Parameter Estimate	Standard Error
Model	5			M	6.61	0.34
Age	2	10.42	0.0054	A_1	−0.27	0.64
Number	1	8.82	0.0030	A_2	1.26	0.41
Youngest	2	103.71	0.0001	N	−0.83	0.28
Residual	3	3.43	0.3299	Y_1	4.52	0.45
Total	8			Y_2	−1.00	0.43

The tests of the main effects indicates that each is significant ($p < 0.01$). The parameter estimates indicate the overall mean, adjusted for main effects, is 6.61 physician contacts. This increases by 1.26 contacts if the subscriber is 25–44 and decreases by $(1.26 - 0.27) = 0.99$ contact for subscribers 45–59. Subscribers under 25 are not significantly different from the mean (-0.27 ± 0.64). Thus the effect of age is not uniformly increasing or decreasing.

The number of children has the expected effect of a one-child family having 0.83 fewer than average contacts. The effect of "youngest child" is to increase the average by 4.52 visits under age 5, decrease it by 1 visit for youngest 5–9, and decrease it by $(4.52 - 1.00) = 3.52$ visits for the youngest being at least 10. Here we see that the effect of the youngest child tends to be increasing with decreasing age.

As an aside, the possible linear effect of age can be tested with a contrast matrix as in Section 5.2. Specifically, suppose that the underlying parameters for the three age groups (which are assumed to be of equal width) are denoted $\gamma_1, \gamma_2, \gamma_3$. Then a linear effect corresponds to H_0: $\gamma_1 - \gamma_2 = \gamma_2 - \gamma_3$ or H_0: $\gamma_1 - 2\gamma_2 + \gamma_3 = 0$. The model identifiability conditions correspond to $\gamma_1 = Y_1, \gamma_2 = Y_2, \gamma_3 = -(Y_1 + Y_2)$. Substituting this into the hypothesis yields the reparameterized hypothesis H_0: $Y_1 - 2Y_2 - (Y_1 + Y_2) = -3Y_2 = 0$ or more simply, H_0: $Y_2 = 0$. The corresponding Wald test is $Q = (\hat{Y}_2)^2/V(\hat{Y}_2) = (-1.00)^2/(0.43)^2 = 5.41$ (or 5.50 when more decimal places are used). Hence we reject the hypothesis of a strictly linear effect.

The important issue with incomplete designs is that generally one or more of the interaction terms cannot be estimated. In the case of the pediatric physician data a total of nine such terms had to be left out of the model. The technical term for this is *confounding*. That is, because of the sampling plan the main effects and some of the interaction terms are said to be confounded. To see precisely which terms, we write out the saturated model with a simpler set of identifiability conditions:

$$X_1 = 1$$

$$X_2 = 1 \text{ if } i = 1, \quad 0 \text{ otherwise}$$

$$X_3 = 1 \text{ if } i = 2, \quad 0 \text{ otherwise}$$

$$X_4 = 1 \text{ if } j = 1, \quad 0 \text{ otherwise}$$

$$X_5 = 1 \text{ if } k = 1, \quad 0 \text{ otherwise}$$

$$X_6 = 1 \text{ if } k = 2, \quad 0 \text{ otherwise}$$

$$X_7 = X_2X_4 \quad X_8 = X_3X_4 \quad X_9 = X_2X_5 \quad X_{10} = X_2X_6 \quad X_{11} = X_3X_5$$

$$X_{12} = X_3X_6 \quad X_{13} = X_4X_5 \quad X_{14} = X_4X_6 \quad X_{15} = X_2X_4X_5$$

$$X_{16} = X_2X_4X_6 \quad X_{17} = X_3X_4X_5 \quad X_{18} = X_3X_4X_6.$$

The detailed design matrix is shown in Table 5.13. For obvious reasons, this is called a (1, 0) or *baseline parameterization.*

Examine the effects of the deleted samples: $X_{10} = X_{14} = X_{16} = X_{18} = 0$, so A_1Y_2, NY_2, A_1NY_2 and A_2NY_2 cannot be estimated. Next observe that $X_2 = X_4$, $X_3 = X_{11}$, $X_7 = X_{15}$, and $X_8 = X_{17}$ hence A_1 and A_1Y_1, A_2 and A_2Y_1, A_1N, and A_1NY_1 and A_2NY_1 are indistinguishable. Finally, $X_{13} = X_7 + X_8$, hence NY_1 cannot be estimated if A_1N and A_2N are estimated. Thus in addition to the coefficients of the main effects, $X_1 - X_6$, only X_7, X_8, and X_{12}, corresponding to A_1N, A_2N, and A_2Y_2, remain. Notice that quite arbitrarily the remaining effects have been labeled main effects. This is done by convention, but the analyst must keep in mind that for this design the main effect and interaction are indistinguishable.

It was fortuitous that all the main effects could be estimated. In some studies the pattern of missing samples means that main effects cannot be distinguished. This is termed a *confounded design* and without making substantive decisions about the probable importance of the confounded factors, data from such designs cannot be analyzed.

Table 5.13 Saturated Model for a 3 × 2 × 3 Design with a (1,0) Parameterization

A	N	Y	Design Coefficients																	
i	j	k	X_1	X_2	X_3	X_4	X_5	X_6	X_7	X_8	X_9	X_{10}	X_{11}	X_{12}	X_{13}	X_{14}	X_{15}	X_{16}	X_{17}	X_{18}
1	1	1	1	1	0	1	1	0	1	0	1	0	0	0	1	0	1	0	0	0
1	1	2																		
1	1	3																		
1	2	1	1	1	0	0	1	0	0	0	1	0	0	0	0	0	0	0	0	0
1	2	2																		
1	2	3																		
2	1	1	1	0	1	1	1	0	0	1	0	0	1	0	1	0	0	0	0	0
2	1	2																		
2	1	3																		
2	1	1	1	0	1	0	1	0	0	0	0	0	1	0	0	0	0	0	0	0
2	2	2	1	0	1	0	0	1	0	0	0	0	0	1	0	0	0	0	0	0
2	2	3	1	0	1	0	0	0	0	0	0	0	0	0	0	0	0	0	0	0
3	1	1																		
3	1	2																		
3	1	3	1	0	0	1	0	0	0	0	0	0	0	0	0	0	0	0	0	0
3	2	1																		
3	2	2	1	0	0	0	0	1	0	0	0	0	0	0	0	0	0	0	0	0
3	2	3	1	0	0	0	0	0	0	0	0	0	0	0	0	0	0	0	0	0

5.4 INTERPRETING THE RESULTS: FITTED VALUES AND PREDICTION

The process of model testing, estimation, and model fitting culminates in a set of estimates that can be used for interpreting the study results. As with three-dimensional arrays (Chapter 4), there are three points to cover:

1. Choice of model
2. Tests of significance and parameter estimates
3. Appropriate summary tables

The basic results should be based on a model selected prior to the analysis. Ideally, this model should be selected prior to collecting the data. As noted at several points, the rationale for the model is based on the existing knowledge in the study area. This rationale should be clearly delineated while reporting phase 1 (the hypothesis-testing phase) of the analysis. After testing the fit of the data to the model, residual analysis can be used to evaluate the model assumptions. Clearly, it is not necessary to report the residual analysis in detail, but a brief summary is usually appropriate.

The phase 2 analysis or "model fitting" usually leads to a more refined model for subsequent study. But it is important to emphasize that this phase 2 model is untested and awaits replication with fresh data.

The tests of significance and parameter estimates for either the phase 1 or the phase 2 analysis or both should be reported as has been done in Sections 5.1–5.3. Again it is important that when the phase 2 tables are reported the tests of significance should be labeled as having only nominal rather than true significance levels. Similarly, when phase 2 standard errors are reported, it should be pointed out that these are smaller than would be expected with fresh data. Again this is because the model has been fit to the data.

The final point to be covered is the presentation of appropriate summary tables. Given a set of parameter estimates, applying the design matrix to them generates *fitted values of the response function*:

$$\hat{\mathbf{f}} = \mathbf{X}\mathbf{b}. \tag{5.10}$$

The associated standard errors can be generated from

$$\mathbf{V}(\hat{\mathbf{f}}) = \mathbf{X}(\mathbf{X}'\mathbf{V}^{-1}\mathbf{X})^{-1}\mathbf{X}' \tag{5.11}$$

as was discussed in Appendix 4A.1. Most standard programs generate $\hat{\mathbf{f}}$ and $\mathbf{V}(\mathbf{f})$ as an option. Examples for the data sets discussed in Sections 5.1–5.3 are shown in Tables 5.14–5.16. Notice that these fitted values must be generated for the phase 1 residual analysis, so no additional computations are required except for phase 2 fitted values.

The fitted values for cancer knowledge (phase 1 main effects) show an interesting gradient. The highest level is for people who participate in all four activities (6.95 percent). The next cluster of values is from 54 to 63 percent. These are persons who do solid reading and either read newspapers or do both, listen to the radio, and attend lectures. The third cluster (around 47.5 percent) either does solid reading and only one other activity or all three other activities. The fifth cluster (38.7 to 40.7 percent) either does solid reading or reads newspapers and one of the other two. The next group does no solid reading but either only reads the newspaper or only goes to lectures or listens to the radio. The lowest level, as might be expected, corresponds to people who participate in none of the activities. In general terms, the more one does, the better he or she is informed. However, solid reading is equivalent to any two other activities. Interestingly, attending lectures appears to have little more effect than listening to the radio. But recall that the phase 2 analysis suggested that while the effect of radio applies evenly (main effect only), the lecture effect may depend on either solid reading or newspaper reading. The prevalence of bysinnosis is dramatically affected by working in dusty areas, as shown in Table 5.15. If one does not work in such areas, the prevalence is less than 2 percent. However, in dusty areas the prevalence is at least 4.6 percent. Moreover, it increases after 10 years of employment in dusty areas. Finally, smoking in dusty areas also increases the prevalence. The fitted values for pediatric physician contacts are shown in Table 5.16. Here the effects are most easily interpreted from the parameter estimates (Section 5.3 and Table 5.12). However, the fitted means can be used to display the results graphically, as in Figure 5.1. Two points emerge. First the curvilinear age

Table 5.14 Fitted Percentages for Good Cancer Knowledge by Reading Newspapers, Solid Reading, Listening to Radio, and Attending Lectures (Main-Effects Model)

		Solid Reading:	Yes(1)		No(−1)	
		Attend Lectures:	Yes(1)	No(−1)	Yes(1)	No(−1)
Read newspapers	Yes(1)	Listen to radio Yes(1)	69.5	60.9	47.3	38.7
		No(−1)	62.9	54.3	40.7	32.1
	No(−1)	Yes(1)	55.1	46.5	32.9	24.3
		No(−1)	48.5	39.9	26.3	17.7

Table 5.15 Fitted Prevalence[a] of Byssinosis by Working in Dusty Areas, Years Employed, and Smoking Habit (Modular Model)

Dust Area	Years Worked		Smoking Habit Never	Smoking Habit Ever
Yes(1)	<10	(1,0)	4.6	14.0
	10–<20	(0,1)	14.6	24.1
	20+	(−1,−1)	14.5	23.9
No(−1)	<10	(0,0)	1.2	1.2
	10–<20	(0,0)	1.2	1.2
	20+	(0,0)	1.2	1.2

[a]Computations before rounding parameter estimates.

Table 5.16 Observed and Fitted Mean Pediatric Physician Contacts by Subscriber's Age, Number of Children, and Age of Youngest Child

Age	Number	Youngest	Observed	Fitted
<25	1	<5	9.30	10.03
	2 +	<5	12.51	11.69
25–44	1	<5	10.75	11.57
	2 +	<5	13.40	13.23
	2 +	5–9	7.97	7.71
	2 +	10+	4.65	5.18
45–59	1	10+	1.52	1.27
	2 +	5–9	4.73	5.46
	2 +	10+	2.62	2.93

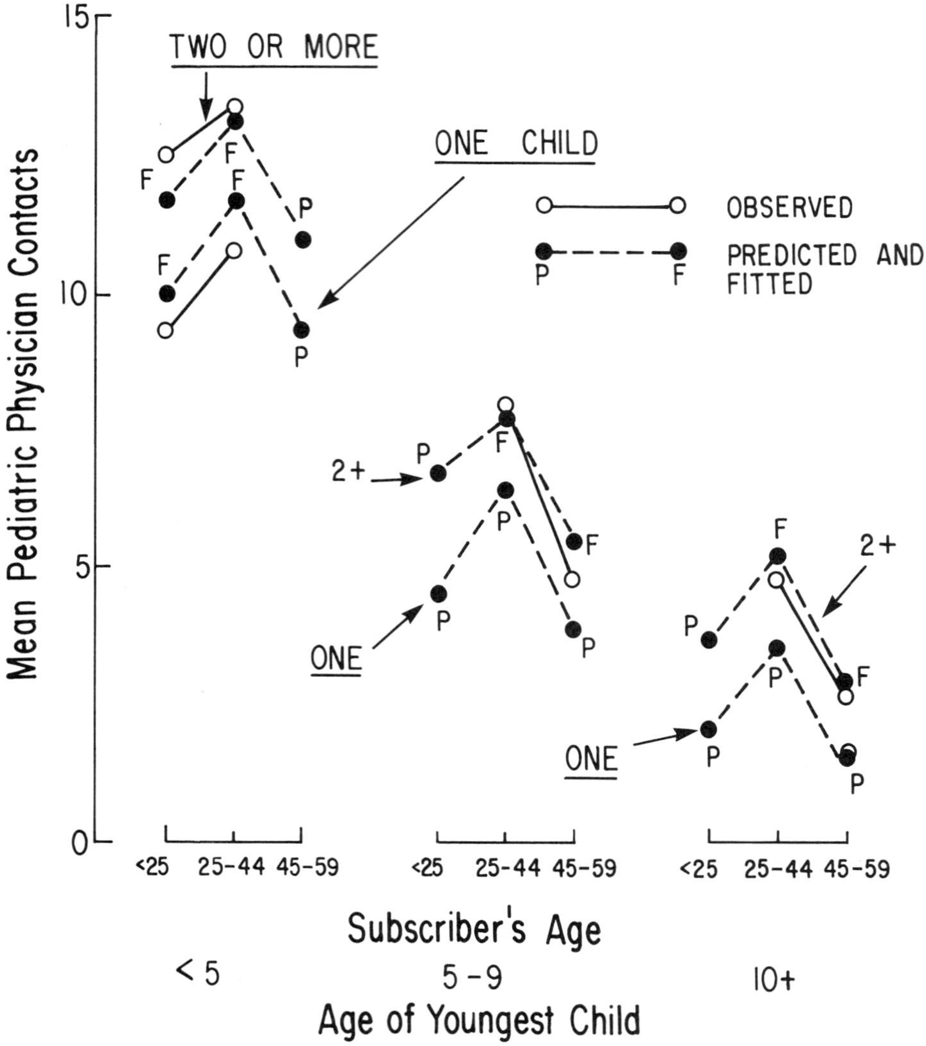

Figure 5.1 Observed, fitted, and predicted mean pediatric physician contacts.

effect and the uniformly decreasing youngest child effects are now apparent, as is the family-size effect. Second, the relatively sparse nature of the sampling is also evident from there being only 9 rather than 18 points plotted.

A second use of the fitted means is for planning purposes. Suppose that a new group is being added to the HMO and that only the age distribution of the subscribers is known. However, the average family size is two children and the average age of the youngest child is 5–9. Suppose that the age distribution is 100 <25, 300 25–44, and 100 45–59. Using the average ages we would project pediatric contacts as

$$
\begin{aligned}
&100(6.61 - 0.27 + 0.83 - 1.00) + 300(6.61 + 1.26 + 0.83 - 1.00) \\
&\qquad + 100(6.61 - (-0.27 + 1.26) + 0.83 - 1.00) \\
&= 100(6.17) + 300(7.7) + 100(5.45) = 3472.
\end{aligned}
$$

This must be considered as only a rough guide for two reasons. First, only averages for number of children and youngest were used. The actual distributions could distort this considerably. Second, and perhaps more crucial, for subscribers under 25 no samples had children 5–9; hence we are extrapolating into an unknown region, which is always risky.

6

Multiple-Response Models

The one-response, multifactor models of Chapter 5 assumed that a data set was collected to examine a well-defined set of hypotheses about associations between the factors and the response. Implicitly, it was assumed that the associations among the factors were of little substantive interest or had been specified by the study design. Many studies are much more exploratory in nature. In this circumstance associations among all the variables are of interest. Such studies include the no-factor, multiple-response design and the one-factor multiple-response design.

Two special cases are the two-response, multiple-factor study and the case-control study. The first case is a straightforward extension of the methods of Chapter 5. The association between two responses is usually summarized in terms of a single index of association. The index is then treated as the response and the analysis proceeds as a one-response, multifactor analysis.

The second design is the case-control design where the factor (case or control) should be treated as a response. Because the odds ratio is symmetric, this can be done. This leads to an analysis based on logistic regression, which is discussed in Chapter 7.

The multiple-response analysis is conceptually similar to the analysis of three responses. It is common to utilize the log-linear model and the associated relative odds ratios to explore the dependencies among the variables. The major difference is: When the number of responses is greater than three, the number of possible models increases dramatically. For example, if only four responses are considered, there are 114 models involving four variables, and over 166 models if models involving three or fewer variables are considered. Hence the problem of model selection is much more difficult. In fact, there is no single optimal approach as there was for three variables.

The residual analysis is somewhat more complex as well. The difficulty is primarily one of identifying the appropriate set of residuals for the rankit plots.

The collapsibility theorems extend in a straightforward manner to four or more variables. A geometric aid will be introduced to assist in identifying the appropriate summary tables. This aid can also be used to guide our interpretation of the results. As noted by Fienberg (1980), when the variables are all dichotomous, the aid can be utilized in a manner analogous to path analysis (Sewall 1930).

6.1 THE SATURATED LOG-LINEAR MODEL

The data in Table 6.1 were reported by Wermuth (1976) in a discussion of techniques for model selection. They show the distribution of birth outcomes by age of mother, length of gestation, and number of cigarettes the mother smoked per day during the prenatal period. Wermuth gives a more detailed discussion of the rationale for the various cut points.

The first step is to extend the log-linear model to more than three dimensions. Let the probability that a woman age h, smoking status i, has an infant of gestational age j with birth outcome k be denoted $P\{A = h, S = i, G = j, D = k\} = p_{hijk} > 0$, where $\Sigma_h \Sigma_i \Sigma_j \Sigma_k p_{hijk} = 1$. The corresponding random variable is denoted Y_{hijk}, which under the usual assumptions of simple random sampling will follow a multinomial distribution for a sample of size n. The expected value of Y_{hijk} is $m_{hijk} = E(Y_{hijk}) = np_{hijk}$.

Table 6.1 Infant Survival and Gestational Age According to Mother's Age and Smoking Habits

		Gestational Age in Days						
		197–260		261+		Total		
		Infant Survival at One Year						
Mother's Age	Amount Smoked	No $(hi11)$	Yes $(hi12)$	No $(hi21)$	Yes $(hi22)$	No $(hi+1)$	Yes $(hi+2)$	Total $(hi++)$
<30	<5 $(1, 1, j,k)$	50	315	24	4012	74	4327	4401
	5+ $(1, 2, j,k)$	9	40	6	459	15	499	514
	Total $(1, +, j,k)$	59	355	30	4471	89	4826	4915
30+	<5 $(2, 1, j,k)$	41	147	14	1594	55	1741	1796
	5+ $(2, 2, j,k)$	4	11	1	124	5	135	140
	Total $(2, +, j,k)$	45	158	15	1718	60	1876	1936
Total	<5 $(+, 1, j,k)$	91	462	38	5606	129	6068	6197
	5+ $(+, 2, j,k)$	13	51	7	583	20	634	654
	Total $(+, +, j,k)$	104	513	45	6189	149	6702	6851

Source: Wermuth (1976). Reprinted with permission from The Biometric Society.

Let $l_{hijk} = \ln(np_{hijk}) = \ln(m_{hijk})$. Since adding the fourth variable, A, is entirely analogous to adding a third variable as in Chapter 4, the saturated log-linear model is immediate:

$$\begin{aligned} l_{hijk} = m &+ A_h + S_i + G_j + D_k + AS_{hi} + AG_{hj} + AD_{hk} + SG_{ij} + SD_{ik} \\ &+ GD_{jk} + ASG_{hij} + ASD_{hik} + AGD_{hjk} + SGD_{ijk} \\ &+ ASGD_{hijk}, \end{aligned} \tag{6.1}$$

where the parameters are defined as deviations from the corresponding lower-order means as in (4.54). This of course implies the usual identifiability conditions: $\Sigma_h A_h = \Sigma_i S_i = \Sigma_j G_j = \Sigma_k D_k = \Sigma_h AS_{hi} = \cdots = \Sigma_k ASGD_{hijk} = 0$. The parameter maximum likelihood estimates may be obtained by replacing the m_{hijk} by the observed counts y_{hijk}. For example,

$$\hat{A}_1 = -\hat{A}_2 = \frac{1}{16} \ln \left[\frac{50(315)(24)(4012)(9)(40)(6)(459)}{41(147)(14)(1594)(4)(11)(1)(124)} \right] \Big/ 16,$$

or $A_1 = 0.477$. The antilog of A_1 equals the estimate that a randomly selected woman is about $\exp(2A_1) = 2.59$ times as likely to be under 30 as 30 or older, after adjusting for all other effects. The estimate is based on the fact that for a dichotomous variable, main effects are half the corresponding log odds.

A similar computation yields the estimates for the association between gestation and survival:

$$\begin{aligned} \hat{GD}_{11} &= -\hat{GD}_{12} = -\hat{GD}_{21} = \hat{GD}_{22} \\ &= \frac{1}{16} \ln \left\{ \left[\frac{50(4012)}{315(24)} \right] \left[\frac{9(459)}{40(6)} \right] \left[\frac{41(1594)}{147(14)} \right] \left[\frac{4(124)}{1(11)} \right] \right\} \\ &= \ln \frac{[26.534(17.212)(31.756)(45.091)]}{16} = 0.837. \end{aligned}$$

These and the remaining parameter estimates are shown in Table 6.2. The corresponding approximate standard error for all estimates in the saturated model is $SE = (\Sigma_h \Sigma_i \Sigma_j \Sigma_k 1/y_{hijk})^{1/2}/16 = 0.084$. The ratios of the estimate to the standard error, or Wald statistics, are also shown. As can be seen, only AS and GD are nominally significant. This clearly suggests that a simpler model may be appropriate.

Table 6.2 Log-Linear Parameters for Wermuth's Data

Label	Saturated Model		[AD,DG,DS,SA,AG]		[AS,AD,GD]	
(1,1,1,1)	Estimate	z-Test	Estimate	z-Test	Estimate	z-Test
M	3.89	—	3.91	—	3.83	—
D	−1.54	−18.25	−1.51	−22.99	−1.59	−33.83
S	1.09	12.92	1.07	17.23	1.18	47.29
A	0.48	5.65	0.41	8.62	0.42	8.86
G	−0.39	−4.59	−0.40	−8.58	−0.41	−8.97
DS	−0.08	−0.97	−0.11	−1.81	—	—
DA	−0.06	−0.70	−0.12	−2.58	−0.14	−3.25
DG	0.84	9.20	0.83	17.94	0.83	18.06
AS	−0.17	−2.06	−0.10	−4.13	−0.10	−4.07
SG	−0.00	−0.04	—	—	—	—
AG	−0.09	−1.11	−0.04	−1.72		
DSA	−0.06	−0.70				
DSG	0.01	0.06				
DAG	−0.07	−0.85				
SAG	0.03	0.37				
DSAG	0.05	0.58				

6.2 MODEL SELECTION

The problem of model selection when only three variables were under consideration is reasonably straightforward. All possible models are estimated and the log-linear parameters are deleted in a sequential fashion until a model is identified. In addition, since all models are estimated, any lower-order models which are not arrived at by the backwards elimination of terms can also be examined. Because there are so many possible models when four or more variables are considered, this is not a feasible approach.

There are a number of algorithms or procedures that attempt to arrive at an appropriate model without considering all possible models. Two of these, "forward selection" and "backwards elimination," are analogous to what is done in multiple regression analysis. These have the disadvantage of sometimes being computationally formidable. A third approach, due to Brown (1976) attempts to identify an appropriate model by considering "partial associations." A fourth approach is based on the findings of the tests of significance in the saturated model. An example of this approach was considered in Section 6.1. We will examine each of these for Wermuth's data with four responses. Several additional algorithms are found in Bishop et al. (1975, Chap. 4).

It is important to recognize that none of the algorithms is guaranteed to find the best possible model for a data set. In fact, different algorithms will sometimes yield different models. The analyst can only rely on his or her judgment in deciding which model is the one that is most appropriate for summarizing the associations in the data. This judgment is guided not only by the automatic results of the algorithm but also residual analysis and the original objectives of the study. This last point is to remind the analyst that if the objective was at least partly to study the association between A and B, it makes no sense to collapse over A or B.

The algorithm that most closely resembles the process for three variables is *backwards elimination*. In this algorithm for s variables the model with $s(s - 1)$ variable associations is fit and each of these terms is deleted in turn. The model that produces the smallest increase in the residual chi-square is then selected. If there are differing degrees of freedom, the least significant term is dropped first. The process continues until all the remaining terms would increase the residual chi-square by a significant amount.

The steps of this algorithm are shown in Table 6.3 for the data discussed in Section 6.1. It begins by fitting the model of all four three-variable interactions. The residual chi-square is quite small ($G^2 = 0.36$), so we conclude that there is little evidence of a four-variable interaction. The first step is to fit the four possible models with three three-variable interactions. The deleted terms are indicated by an asterisk and the fit statistics and corresponding change in fit are again displayed. The model that excludes ASG has the smallest overall fit statistic ($G^2 = 0.41$) and the smallest change in fit statistic ($G^2 = 0.05$), so it is selected at step 1.

Normally, the algorithm would proceed by deleting each of the three-variable terms and examining the corresponding model chi-squares. However, the process can be accelerated by noting that each

Table 6.3 Backwards Elimination for Wermuth's Data on Infant Survival (Table 6.1)

Model[a]	df	G^2	Δdf	ΔG^2	Selected Model
[*ASG*, *ASD*, *AGD*, *SGD*]	1	0.36	—	—	
[*ASG*, *ASD*, *AGD*, *]	2	0.58	1	0.23	
[*ASG*, *ASD*, * , *SGD*]	2	0.81	1	0.45	
[*ASG*, * , *AGD*, *SGD*]	2	0.69	1	0.33	
[* , *ASD*, *AGD*, *SGD*]	2	0.41	1	0.05	1
. . .	. . .	. . .	. . .	. . .	. . .
[*AS*, *AG*, *AD*, *SG*, *SD*, *GD*]	5	1.72	4	1.36	2
[*AS*, *AG*, *AD*, *SG*, *SD*, *]	6	339.32	1	337.60	
[*AS*, *AG*, *AD*, *SG*, * , *GD*]	6	4.04	1	2.32	
[*AS*, *AG*, *AD*, * , *SD*, *GD*]	6	1.82	1	0.10	3
[*AS*, *AG*, * , *SG*, *SD*, *GD*]	6	8.17	1	6.45	
[*AS*, * , *AD*, *SG*, *SD*, *GD*]	6	4.69	1	2.97	
[* , *AG*, *AD*, *SG*, *SD*, *GD*]	6	19.96	1	18.24	
[*AS*, *AG*, *AD*, *SD*, *GD*]	6	1.82	—	—	
[*AS*, *AG*, *AD*, *SD*, *]	7	340.06	1	338.24	
[*AS*, *AG*, *AD*, * , *GD*]	7	4.79	1	2.97	
[*AS*, *AG*, * , *SD*, *GD*]	7	8.30	1	6.48	
[*AS*, * , *AD*, *SD*, *GD*]	7	4.74	1	2.92	4
[* , *AG*, *AD*, *SD*, *GD*]	7	20.01	1	18.19	
[*AS*, *AD*, *SD*, *GD*]	7	4.74	—	—	
[*AS*, *AD*, *SD*, *G*]	8	347.07	1	342.33	
[*AS*, *AD*, * , *GD*]	8	7.71	1	2.97	5
[*AS*, * , *SD*, *GD*]	8	15.46	1	10.72	
[* , *AD*, *SD*, *GD*]	8	22.93	1	18.19	
[*AS*, *AD*, *GD*]	8	7.71	—	—	—
[*AS*, *AD*, *G*]	9	350.05	1	342.34	
[*AS*, * , *GD*]	9	17.84	1	10.13	
[*S*, *AD*, *GD*]	9	25.31	1	17.60	

Total models: 27

[a]Asterisk denotes deleted term.

of the three-variable terms produces a change in chi-square of less than 1 (df = 1 for each). This suggests immediately fitting the model with no three- or four-variable interactions. This is the model with six two-variable interactions or all partial associations. Since the change in fit statistic ($\Delta G^2 = 1.36$, df = 4) would not be significant for any single term, this model is selected at step 2.

Step 3 proceeds by fitting the six models with one two-variable interaction, or partial association, removed. The deletion of *SG* produces the smallest change in fit statistic ($\Delta G^2 = 0.10$), so the model [*AS, AG, AD, SD, GD*] is selected at step 3.

Using the selected model, each of the five partial associations is again deleted in turn. This time two candidates yield a similar change in fit statistics:

1. [*AS, AG, AD, GD*]|*SD* deleted ($\Delta G^2 = 2.97$).
2. [*AS, AD, SD, GD*]|*AG* deleted ($\Delta G^2 = 2.92$).

The overall fit of these two models each is quite small ($p > 0.68$), but the change in chi-square suggests that *SD* and *GD* are statistically significant at the 0.10 level. However, following the algorithm, the second model is selected.

Each of the remaining partial associations is deleted from the model selected in step 4: [*AS, AD, SD, GD*]. Note that when *GD* is deleted, the main effect for gestation, *G*, must be retained. At this point the change induced by deleting *SD* is the smallest ($\Delta G^2 = 2.97$), so [*AS, AD, GD*] is selected. If we compare the overall fit of this model ($G^2 = 7.71$, df = 8) to the model selected at step 3 ($G^2 = 1.82$, df = 6), the difference ($\Delta G^2 = 5.89$, df = 2) is the simultaneous test of H_0: $AG = SD = 0$. The corresponding significance level is $0.05 < p < 0.10$, which emphasizes the borderline significance of these two terms. A second point is that the two tests based on deletion add exactly to the joint test. This occurs because *SD* is collapsible with respect to *G* and *AG* is collapsible with respect to *S* in the model selected at step 3. Notice in this model that *SG* has been set equal to zero, so smoking and gestation are conditionally independent. This is an application of the collapsibility theorems to be discussed in Section 6.4.

The algorithm proceeds to examine each of the lower-order relatives of [*AS, AD, GD*]. The candidate model would be [*AS, GD*], but the overall fit statistic is significant at the 0.05 level ($G^2 = 17.84$, df = 9), so the algorithm terminates with the model selected at step 5. It happens that this is the same model selected by examining the saturated model.

A final step, which is not strictly part of the backwards elimination algorithm, is to compare all the neighboring models to the selected model. As can be seen, none of the added terms significantly improve the fit, especially *ADG* and *ASD*, and deleting terms will produce a significant lack of fit (Table 6.4).

The reverse process is known as *forward selection*. Here the algorithm begins with a simple model, say main effects only, and after fitting all the higher-order relatives selects the model that produces the largest reduction in the error chi-square statistic. Here ''largest'' corresponds to most significant or smallest p-value. The algorithm terminates when no additional terms produce a significant reduction in the fit statistic or when the overall fit statistic is nonsignificant. The choice of decision rules is up to the judgment of the analyst.

The forward selection algorithm is applied to Wermuth's data in Table 6.5. The baseline model has only the four main effects [*A*, *S*, *G*, *D*]. This model does not fit the data, $G^2 = 377.76$,

Table 6.4 Neighboring Models to [*AS*, *AD*, *GD*] Selected by Backwards Elimination

Model	df	G^2	Δdf	ΔG^2
		Added Terms		
[*AS*, *AD*, *GD*, *SG*]	7	7.11	1	0.60
[*AS*, *AD*, *GD*, *AG*]	7	4.79	1	2.92
[*AS*, *AD*, *GD*, *SD*]	7	4.74	1	2.97
[*AS*, *AD*, *GD*]	8	7.71	—	—
		Deleted Terms		
[*AS*, *AD*, *G*]	9	350.05	1	342.34
[*AS*, * , *GD*]	9	17.84	1	10.13
[*S* , *AD*, *GD*]	9	25.31	1	17.60

Table 6.5 Forward Selection for Wermuth's Data on Infant Survival

Model	df	G^2	Added Term	Δdf	ΔG^2	Selected Model
[*A* , *S* , *G* , *D*]	11	377.76	—	—	—	—
[*AS*, *G* , *D*]	10	360.17	*AS*	1	17.60	
[*AG*, *S* , *D*]	10	370.74	*AG*	1	7.02	
[*AD*, *G* , *D*]	10	376.63	*AD*	1	10.14	
[*A* , *SG*, *D*]	10	377.25	*SG*	1	0.52	
[*A* , *SD*, *G*]	10	375.38	*DS*	1	2.39	
[*A* , *S* , *GD*]	10	35.45	*GD*	1	342.32	1
[*A* , *S* , *GD*]	10	34.45	—	—	—	
[*AS*, *GD*]	9	17.83	*AS*	1	17.62	2
[*AG*, *GD*]	9	28.43	*AG*	1	7.02	
[*AD*, *S* , *GD*]	9	25.31	*AD*	1	10.13	
[*A* , *SG*, *GD*]	9	34.91	*SG*	1	8.54	
[*A* , *SD*, *GD*]	9	33.06	*SD*	1	2.39	
[*AS*, *GD*]	9	17.83	—	—	—	
[*AS*, *AG*, *GD*]	8	10.82	*AG*	1	7.00	
[*AS*, *AD*, *GD*]	8	7.71	*AD*	1	10.13	3
[*AS*, *SG*, *GD*]	8	17.32	*SG*	1	0.51	
[*AS*, *SD*, *GD*]	8	15.45	*SD*	1	2.38	
[*AS*, *AD*, *GD*]	8	7.71	—	—	—	
[*AS*, *AG*, *AD*, *GD*]	7	4.79	*AG*	1	2.92	
[*AS*, *AD*, *SG*, *GD*]	7	7.11	*SG*	1	0.60	
[*AS*, *AD*, *SD*, *GD*]	7	4.74	*SD*	1	2.97	4

Total models: 22

df = 11. In the first step each of the partial associations is fit to the model in turn. The addition of *GD* reduces the fit statistic the most ($\Delta G^2 = 342.32$, df = 1), so it is selected.

In step 2, each of the five remaining partial terms is added to the current baseline model: [*A*, *S*, *G*, *D*]. This time the largest reduction occurs for the model: [*AS*, *GD*] ($\Delta G^2 = 17.62$, df = 1), so it is selected in step 2 and becomes the baseline for step 3. The resultant model, [*AS*, *AD*, *GD*], induces the greatest reduction ($\Delta G^2 = 10.13$, df = 1). Moreover, it is the first model that can be said to fit the data, $G^2 = 7.71$, df = 8.

At the next step *AG, SG,* and *SD* are considered in turn. None of these three partial associations produces a significant reduction in the fit statistic, so the algorithm terminates with the model selected in step 3. Notice that if the stopping rule had used a nominal significance of 0.10, both *AG* and *SD* would have been selected.

The model identified by forward selection is the same as with backwards elimination and the saturated model approach, so the final step can be summarized as in Table 6.4.

A different approach to model selection is based on a slightly different definition of partial association. The approach is due to Brown (1976) and attempts to identify a model in a single step. We term a model with all interactions involving k variables, a uniform kth-order model. Thus a model for four variables with all four three-variable interactions is a uniform third-order model. Brown defines the partial association test as the change in fit when one interaction is deleted from the corresponding uniform model. The algorithm proceeds by computing all such partial association tests. The identified models consist of all significant partial associations and their lower-order relatives. The second, and usually final step, is to compute all higher- and lower-order relatives of the identified model. If any of the relatives are significant or nonsignificant depending on the direction, either forward selection or backwards elimination is implemented using the identified model as a baseline.

Brown's algorithm is applied in Table 6.6. As can be seen, the uniform second-order model fits quite well, so none of the three-variable interactions need to be examined. The partial association tests indicate that *DG, AD,* and *AS* are significant. Since these include the four main effects, the identified model is [*AS, AD, GD*]. Again, Table 6.4 summarizes the significance of the higher- and lower-order relatives.

For these data all three algorithms identified the same model, but this is not generally true. When different models, usually not hierarchical relatives, are identified, the analyst must rely on substantive knowledge to choose between the models. Sometimes residual analysis, as discussed in Section 6.3, can be used as a guide. Alternatively, the analyst can present all the identified models and the associated param-

Table 6.6 Tests of Partial Association for Wermuth's Data

Model	df	G^2	Deleted Term	Partial Association Δdf	Partial Association ΔG^2
[*A*, *S*, *G*, *D*]	11	377.76	—	—	—
[*A*, *S*, *G*, *]	12	8439.71	*D*	1	8061.95
[*A*, *S*, * , *D*]	12	5727.99	*G*	1	5350.23
[*A*, * , *G*, *D*]	12	5559.22	*S*	1	5181.46
[* , *S*, *G*, *D*]	12	1717.37	*A*	1	1339.61
[*AS*, *AG*, *AD*, *SG*, *SD*, *DG*]	5	1.72	—	—	—
[*AS*, *AG*, *AD*, *SG*, *SD*, *]	6	339.32	*DG*	1	337.60
[*AS*, *AG*, *AD*, *SG*, * , *DG*]	6	4.04	*SD*	1	2.32
[*AS*, *AG*, *AD*, * , *SD*, *GD*]	6	1.81	*SG*	1	0.09
[*AS*, *AG*, * , *SG*, *SD*, *DG*]	6	8.17	*AD*	1	6.45
[*AS*, * , *AD*, *SG*, *SD*, *DG*]	6	4.70	*AG*	1	2.98
[* , *AG*, *AD*, *SG*, *SD*, *DG*]	6	19.96	*AS*	1	18.24
[*ASG*, *ASD*, *AGD*, *SGD*]	1	0.36	—	—	—
[*ASG*, *ASD*, *AGD*, *]	2	0.59	*SGD*	1	0.23
[*ASG*, *ASD*, * , *SGD*]	2	0.81	*AGD*	1	0.45
[*ASG*, * , *AGD*, *SGD*]	2	0.69	*ASD*	1	0.33
[* , *ASD*, *AGD*, *SGD*]	2	0.41	*ASG*	1	0.05

Total models: 17

eter estimates. In the discussion, he or she must attempt to resolve any paradoxes presented by the competing models.

One last point concerns the relative efficiency or speed of the algorithms. Clearly, using the parameter estimates of the saturated model requires the fewest computations. Unfortunately, frequently, this leads to identifying an inappropriate model. In the present case the partial association algorithm identified a model after examining 17 candidates. For five variables it requires 34, and in general for five variables $2^s + s - 3$ models. Generally, because of the more restricted search, forward selection and backward elimination require fewer models. This can be reinforced by beginning backward elimination with the model of uniform lowest order that fits and forward selection with the model of uniform highest order that does not fit. For example, with Wermuth's data backwards elimination would have fit 23 rather than 27 models. Finally, when forward selection begins with main effects only, many of the initial models can be fit with a single cycle of the IPF procedure. This also tends to make it computationally simpler than the other search procedures. Hence it is especially useful for the initial exploration of large and possibly complex contingency tables.

6.3 RESIDUAL ANALYSIS

The assumptions of our analysis are that the observations are independently and identically distributed. This leads to the counts following a multinomial distribution. The likelihood ratio chi-square statistics, which guided the model selection, also require that the sample be sufficiently large to ensure the appropriate sampling distribution. These assumptions can be examined using residual analysis in the same manner as with three variables.

Residual analysis addresses two questions:

1. Are there any unusually large residuals?
2. Is the pattern consistent with normality?

As before, the first question requires examining all the residuals, and the second requires adjusting for the tables fixed by the model.

The observed and fitted values, together with the adjusted residuals, are shown in columns (1)–(3) of Table 6.7 for the data in Table 6.1. Here the selected model is [*AS*, *AD*, *GD*]. The smallest adjusted residual is -0.87 and the largest is 1.90; certainly this does not suggest any outliers.

The next question concerns the overall pattern. The model [*AS*, *AD*, *GD*] fixes m_{hi++}, m_{h++k}, and m_{++jk}. One pattern of deleted cells that accounts for this is indicated by the "-" in Table 6.7. This was constructed by starting with the (1, 1, 1, 1) cell and incrementing the index values from right to left until one of the fixed tables occurred. Notice that the entire set m_{2ijk} has been deleted.

The next step is to construct the rankits. The procedure is illustrated in columns (4)–(6). Notice there are only eight adjusted residuals to be considered, corresponding to the eight degrees of freedom

Table 6.7 Residual Analysis of Table 6.1 Using *AS*, *AD*, *GD*[a]

Age A h	Smoking S i	Gestation G j	Survival D k	Observed y (1)	Fitted m (2)	Adjusted Residual z (3)	Rank h (4)	Adjusted Rank $\frac{3h'-1}{3(8)+1}$ (5)	Rankit (6)
1	1	1	1	50	55.62	−0.75	2	0.20	−0.84
1	1	1	2	315	330.77	−0.87	1	0.08	−1.41
1	1	2	1	24	24.07	−0.01	4	0.44	−0.15
1	1	2	2	4012	3990.54	0.34	—	—	—
1	2	1	1	9	6.50	0.98	7	0.80	0.88
1	2	1	2	40	38.63	0.22	5	0.56	0.18
1	2	2	1	6	2.81	1.90	—	—	—
1	2	2	2	459	466.06	−0.33	—	—	—
2	1	1	1	41	38.85	0.34	6	0.68	0.50
2	1	1	2	147	133.21	1.20	8	0.92	1.48
2	1	2	1	14	16.81	−0.68	3	0.32	−0.47
2	1	2	2	1594	1607.13	−0.33	—	—	—
2	2	1	1	4	3.03	1.74	—	—	—
2	2	1	2	11	10.38	0.19	—	—	—
2	2	2	1	1	1.31	−0.27	—	—	—
2	2	2	2	124	125.28	−0.11	—	—	—

[a]—, Cells deleted because of fixed margins: m_{hi++}, m_{h++k}, m_{++jk}.

of the residual chi-square. The plot of the adjusted residuals versus the rankits is shown in Figure 6.1, together with the assumed normal line. There are four points above and below the line. Moreover, the crossing appears to be haphazard. We conclude that there is little evidence against the large-sample assumption.

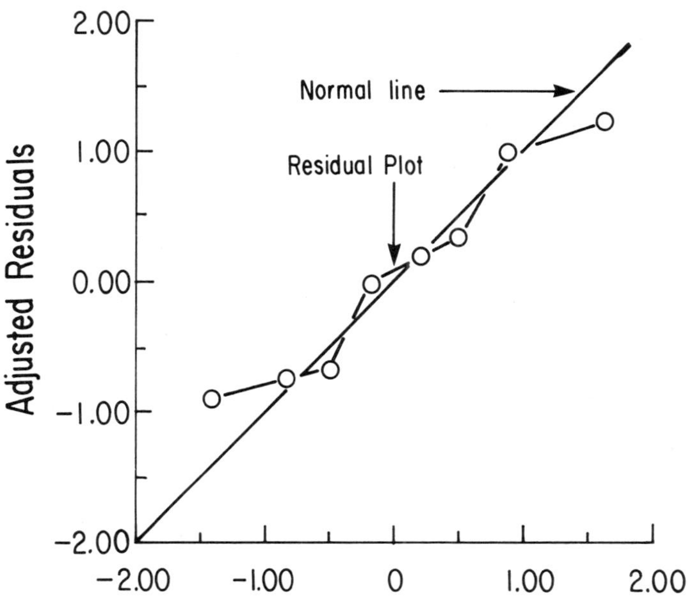

Figure 6.1 Rankit plot for data on infant mortality (Table 6.1): Mode. [*AS*, *AD*, *GD*].

6.4 COLLAPSIBILITY

The process of identifying a model for summarizing the data leads to one of two possibilities:

1. All partial associations and possibly one or more hierarchical relatives are significant.
2. At least one partial association and its hierarchical relatives are nonsignificant.

The first possibility means that the complete multiway table should be reported and the interpretation of data will need to be based on the estimated model. This occurs for the same reason as in three-response tables: A partial association will be collapsible with respect to a third variable when one of the variables in the partial association is conditionally independent of the third variable. The exception to this rule occurs when the table is what Darroch (1976) and Whittemore (1978) refer to as "perfect tables."

When at least one partial association is zero, one or more of the remaining partial associations (and higher-order relatives) will be collapsible with respect to the variables of the zero partial association. In addition, partial association can usually be tested using direct estimators. That is, iterative methods need not be employed. The procedure for these direct tests is discussed in detail by Bishop et al. (1975, Chap. 3, Sec. 4). Fortunately, the iterative methods will converge in a single cycle when these direct estimates exist. What is of greater interest is the ability of the analyst to collapse the data.

To illustrate table collapsing, consider the model with five partial associations for the relationship between age, smoking, gestation, and infant death in Table 6.1: [*AS, AG, AD, SG, SD*]. During the model selection process, it was found that this model fits quite well, $G^2 = 1.82$, df = 6, so all of the three-way and four-way interactions, as well as the smoking gestation partial association, were declared nonsignificant.

The model terms can be arrayed as in Table 6.8. Column (4) shows the *SG* partial association and its higher-order relatives. The fitted model implies that all of these are zero and hence smoking and gestation are said to be conditionally independent. Under the sufficient conditions for collapsibility, this implies that the associations in columns (2) and (3) are collapsible with respect to the excluded variable.

Table 6.8 Model Effects for Infant Data of Table 6.1[a]

Terms involving:			
Neither *S* nor *G* (1)	*S* and Not *G* (2)	*G* and Not *S* (3)	*G* and *S* (4)
M	*S*	*G*	*SG*
A	*AS*	*AG*	*ASG*
D	*SD*	*GD*	*SGD*
AD	*ASD*	*AGD*	*ASGD*

[a]Solid indicates terms not significant ($p > 0.10$).

For example, the gestation survival partial association can be validly estimated from the three-variable table defined by age, gestation, and infant survival. Using the definitions for three-variable tables (4.54):

$$GD^{(3)}_{111} = \frac{1}{8} \ln \left\{ \left[\frac{59(4471)}{30(355)} \right] \left[\frac{45(1718)}{15(158)} \right] \right\} = \ln [24.769(32.620)]$$
$$= 0.837,$$

which is identical to the value shown for the two reduced models in Table 6.2. Here the superscript indicates the three-variable table was used.

This deletion of a single partial association (*SG*) has the important ramification that four of the remaining five partial associations can be estimated from the corresponding three-way tables. The exception is the *AD* partial association.

At the next step of model selection, the *SD* term was deleted. This means that *AD* can be validly collapsed over *S*. Finally, *AG* was deleted, so *AD* can be collapsed over *G* as well.

The selected model is [*AD*, *AS*, *GD*], which corresponds to accepting the hypothesis $AG = SD = SG = 0$ as well as the higher-order relatives. Which tables need to be reported? First, consider the significant associations: AD may be collapsed over S and G since $SD = AG = 0$. Similarly, AS may be collapsed over G and D since $AG = SD = 0$. For the third association, GD, it too can be collapsed over A and S since $AG = SD = 0$. So effectively, these three associations can be reported with three 2×2 tables because of the simple result $AG = SD = 0$.

Next consider the nonsignificant associations:

1. AG may be collapsed over S but not D, since $SG = 0$, but neither AD nor GD is zero.
2. SD may be collapsed over G, but not A, since $SG = 0$, but neither AD nor AS is zero.
3. SG may be collapsed over A or D, since $AG = SD = 0$.

Hence SG requires either ASG or SGD, but AG requires the AGD table and SD requires the ASD table. Putting this altogether, the partial associations in Table 6.1 may be validly summarized with three tables:

1. SG requires age by sex by gestation or sex by gestation by survival.
2. AG, AD, and GD require age by gestation by survival.
3. AS, AD, and SD require age by smoking by survival.

Notice that AD can be obtained from either table 2 or 3.

For complex tales, especially in five or more dimensions, it is sometimes difficult to ascertain the patterns for collapsing tables. A geometric aid has been suggested by Freeman and Jekel (1980). They represent variables by points on a plane, partial associations by straight lines, and triple associations by triangles. The models for Wermuth's data are shown in Figure 6.2. Here dashed lines indicate partial associations significant at the 0.10 level and solid lines at the 0.05 level.

First, consider the model with $SG = 0$. This corresponds to both the solid and dashed lines. Collapsing over smoking moves the point indicated by smoking into the line connecting A and D. Hence AD is not collapsible with respect to smoking, but AG and GD are. Similarly, collapsing gestation affects AD, but neither AS nor SD.

Second, when the dashed lines are ignored, the selected model, AS, AD, GD, emerges. Here it is evident that each significant partial association can be collapsed over the remaining two variables. The SD and AG associations can only be collapsed across G and S, respec-

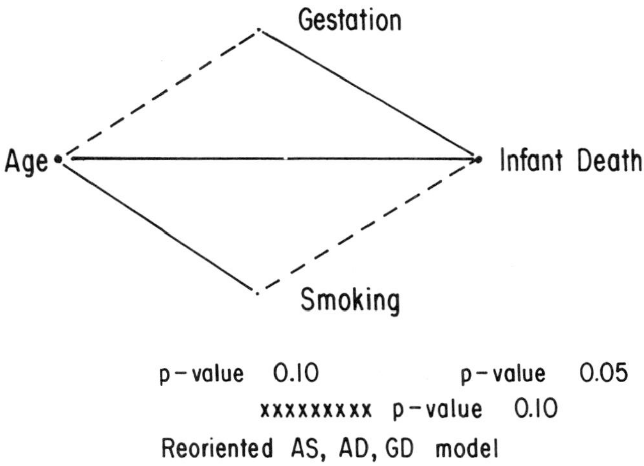

Figure 6.2 Geometric representation of [*AS*, *AG*, *AD*, *SD*, *GD*] and [*AS*, *AD*, *GD*] models for Wermuth's data.

tively. Finally, *SG* is collapsible with respect to *A* or *D*, but not both, because it corresponds to the reoriented diagram at the bottom of Figure 6.2. Thus if we collapse *D*, the *AG* association is affected; hence *SG* cannot also be collapsed over *A*. Conversely, if we collapse *SG* over *A*, the *SD* association is affected, so *SG* cannot also be collapsed over *AD*.

6.5 INTERPRETING THE ANALYSIS

The process of model selection and residual analysis culminates by identifying one or more models for summarizing the associations in a multidimensional contingency table. There are three methods for reporting the results: (1) estimated model and tests of significance; (2) fitted values and odds ratios; (3) collapsed tables. Each of these has advantages and disadvantages in terms of communicability.

The estimated parameters and tests of significance for Wermuth's data are shown in Tables 6.2 and 6.4. From the statistical tests for the model [*AD, AS, GD*], we conclude that gestation is independent of age and smoking. Moreover, infant survival is also independent of the smoking classification. The remaining partial associations are significant. Using the model parameters, we conclude that younger mothers and longer gestation are both associated with better survival [*AD, GD*]. Similarly, younger mothers are likely to smoke 5+ cigarettes. Finally, since no higher-order interaction terms are significant, these effects operate "independently." More correctly, we say that the effects are additive on the log-odds scale or multiplicative on the odds scale.

The second approach is to report the fitted values and relative odds ratios. These can be generated from the estimated model parameters. The fitted values for [*AS, AD, GD*] are shown in Table 6.9. Here gestation and survival have been reversed from Table 6.1, so that the fixed margins may be more readily identified. Using this table, the fitted odds ratios may readily be computed.

For example, if we consider the *GD* odds ratio at age <30 and smoking <5, we obtain: *GD* odds ratio = [55.624/24.068]/[330.771/3990.537] = 27.88. For any other age–smoking category we obtain the same value (except for rounding error). Also, notice that if we collapse over smoking the fitted *AGD* margin yields for age <30: *GD* odds ratio = (62.12/26.879)/(369.402/4456.598) = 27.88, which illustrates the collapsibility results, as does the margin when we collapse over age and smoking.

Next consider the smoking–gestation effect:

$$\text{Age} <30\text{–dead:} \quad SG \text{ odds ratio} = \frac{55.624/24.068}{6.496/2.811} = 1.00.$$

$$\text{Dead:} \quad SG \text{ odds ratio} = \frac{94.475/40.878}{9.524/4.121} = 1.00.$$

$$\text{Marginal:} \quad SG \text{ odds ratio} = \left(\frac{94.475 + 463.984}{40.878 + 5597.663}\right)\left(\frac{4.121 + 591.338}{9.524 + 49.015}\right) = 1.01.$$

Notice that using the fitted values, the association collapses across the levels of the remaining two variables. However, using the *SG* marginal table of observed values yields $SG^{(2)}_{11\cdot} = 553(590)/(5644)(64) = 0.90$.

Table 6.9 Fitted Counts for the Model [*AS*, *AD*, *GD*]

Mother's Age	Mother's Smoking	Dead: 197–260	Dead: 261+	*ASD*: Total	Alive: 197–260	Alive: 261+	*ASD*: Total	*AD*: Total
<30	<5	55.624	24.068	79.692	330.771	3990.537	4321.308	4401
	5+	6.496	2.811	9.307	38.631	466.061	504.692	514
AGD:	Total	62.120	26.879	89	369.402	4456.598	4826	4915
30+	<5	38.851	16.810	55.661	133.213	1607.126	1740.339	1796
	5+	3.028	1.310	4.338	10.384	125.277	135.661	140
AGD:	Total	41.879	18.120	60	143.597	1732.403	1876	1936
SD:	Total							
	<5	94.475	40.878	135.353	463.984	5597.663	6061.647	6197
	5+	9.524	4.121	13.645	49.015	591.338	640.353	654
GD:	Total	104	45	149	513	6189	6702	6851

(Column group heading: Infant Survival and Gestation Days[a])

[a]Integers indicate terms fixed by the model.

Clearly, SG cannot be collapsed over both survival and age. This emphasizes that the collapsibility theorems work for estimated values but not necessarily for fixed or hypothesized values unless the corresponding fitted counts are employed.

In a similar manner, the other two significant odds ratios are AS:OR = (4401/514)/(1796/140) = 0.667, AD:OR = (89/4826)/(60/1876) = 0.577. Thus Table 6.9 and the associated odds ratios can be used to summarize the data. Finally, we can use the three three-dimensional tables of Section 6.4. This has previously been illustrated.

The advantages in terms of communicability depend, in part, on the audience. When parameter estimates and tests of hypotheses are employed, the audience must be familiar not only with general concepts of statistics but also with the parameterizations that are commonly employed and their relative arbitrariness. The use of fitted values and odds ratios tends to appeal more to less sophisticated audiences. However, such tables may be somewhat cumbersome. Finally, when appropriate tests indicate that the data may be validly collapsed, simple two- and three-dimensional arrays can be effectively employed. It should be recognized that when either of the latter approaches is employed, the estimated model can still be reconstructed by the more sophisticated audience.

As a general rule the analyst should report:

1. The original data (e.g., Table 6.1).
2. A summary of the process of model identification.
3. Statistical tests to support conclusions (e.g., Table 6.4).
4. An appropriate set of summary estimates (e.g., one column of Table 6.2 or fitted values or appropriate collapsed tables).

Notice that when parameter estimates are presented, they should be accompanied by either standard errors or z-tests together with a precise statement of the identifiability conditions employed. Finally, an appropriate nonstatistical interpretation of the findings should be provided. Here geometric aids can be especially useful.

EXERCISES

1. The data in Table 6.10 show the results of a five-year follow-up study of the effects of specified treatment of elevated diastolic

Table 6.10 Five-Year Mortality Among HDFP Participants with Average DBP 90–<105 mmHg by Pretreatment Status

Baseline BP	End-Organ Damage	Prior Treatment	Stepped Care		Referred Care	
			Dead	Alive	Dead	Alive
90–<95	No	No	36	986	54	969
		Yes	18	261	20	249
	Yes	No	18	87	16	81
		Yes	12	56	17	61
95–<100	No	No	39	893	53	860
		Yes	16	272	12	273
	Yes	No	8	87	10	69
		Yes	6	69	12	52
100–<110	No	No	31	634	44	723
		Yes	13	203	16	189
	Yes	No	22	65	18	57
		Yes	12	59	19	48

Source: Constructed from Tables 1–4 of Hypertension Detection and Follow-up Program Cooperative Group (1982). Reprinted by permission of the New England Journal of Medicine.

blood pressure. At the start, patients were measured in terms of diastolic blood pressure, the presence or absence of end-organ damage, and whether they were under treatment for elevated blood pressure. All patients can be assumed to be selected on the basis of simple random sampling. The subjects were then randomized into either "stepped care," which was the rigidly specified program of treatment group, or "referred care," which meant referral back to the community. After five years the overall mortality of the two groups was compared. It was found to be 5.9 percent in the stepped-care group and 7.4 percent in the referred-care group. Hence it was concluded that the rigid treatment program should be recommended for all groups. For purposes of analysis, assume that all five variables are responses, so that a log-linear model is appropriate.

(a) Identify a model based on the saturated model.
(b) Identify a model using backwards elimination starting with all four variable interactions. If any high-order interactions

(three or four variables) are significant, decide whether to include them. Justify your decision.

(c) Identify a model using forward selection starting with main effects only, with pairwise interactions. Do the identified models differ?

(d) Identify a model using the partial association algorithm.

(e) Based on the process of model identification, which model seems most appropriate?

(f) Which algorithm do you prefer? Why?

(g) Perform a residual analysis for the selected model and at least one other. How does this change the conclusions in part (e)?

(h) Summarize the findings based on the model selected. Specify any associations that are collapsible with respect to one or more of the baseline variables.

(i) Is stepped care to be recommended across all study groups; that is, do you agree with the study group findings? Justify your answer, but assume that all unreported variables do not affect the conclusions.

2. Stern et al. (1977) surveyed women to study the relationship between Pap testing and hysterectomies. Part of the resulting data is reported in Table 6.11. It is of particular interest to know whether there is an association between Pap testing and ethnicity. The data are cross-sectional, so odds ratios are an appropriate indexes of association.

(a) If the data were longitudinal, what variables could be considered factors?

(b) A log-linear model based on the saturated model.

(c) Identify a log-linear model based on backwards elimination beginning with all triple interactions.

(d) Identify a log-linear model based on forward selection beginning with main effects.

(e) Identify a log-linear model based on partial associations.

(f) Based on the model identification algorithm, which model would you report?

(g) Which algorithm do you prefer? Why?

(h) Perform a residual analysis for the selected model and at

Table 6.11 Relation Between Ethnicity and Pap Testing by Income and Age

		Ethnicity and Years Since Last Test (Including Never)					
		White		Black		Spanish	
Income	Age	<2	2+	<2	2+	<2	2+
Middle	35–44	22	3	37	6	36	4
	45–54	23	3	27	4	30	9
	55–64	23	5	13	1	16	4
	65+	21	8	6	4	3	4
Low	35–44	21	10	39	16	15	5
	45–54	14	11	44	16	5	15
	55–64	11	15	24	18	2	4
	65+	13	15	11	25	3	9

Source: E. Stern, M. Misczyncki, S. Greenland, K. Damus, and A. Coulson (1977).

least one other. Does this confirm your conclusion in part (f)?

(i) Summarize the statistical findings, including whether the ethnicity Pap testing associations are collapsible with respect to age or income or both. Ignore possible effects of unreported variables.

3. For human beings the sex ratio at birth has been established to be 105 (= males/females × 100). However, it has been hypothesized that it may be affected by paternal and/or maternal age. Novitsky and Kimball (1958) investigated this question using the data in Table 6.12.
   (a) Identify a model based on the saturated model.
   (b) Identify a model using backwards elimination.
   (c) Identify a model using forward selection.

Table 6.12 Sex of Offspring by Birth Order, Maternal Age, and Paternal Age, United States, 1955

Age of:		Birth Order and Sex of Offspring			
		First		Second or Higher	
Father	Mother	Male	Female	Male	Female
<20	<20	35,219	32,545	6,523	6,348
	20–24	4,039	3,791	1,766	1,769
	25–29	88	96	91	114
	30–34	7	5	25	25
	35–39	4	2	7	8
20–24	<20	90,109	84,384	41,453	39,803
	20–24	116,912	109,850	138,211	131,753
	25–29	7,974	7,944	14,578	13,698
	30–34	439	441	1,360	1,373
	35–39	45	49	214	203
25–29	<20	21,176	19,719	12,948	12,612
	20–24	86,280	81,467	182,977	174,542
	25–29	48,999	46,188	200,798	191,813
	30–34	5,388	4,951	23,857	22,710
	35–39	562	510	2,588	2,487
30–34	<20	3,122	3,160	2,484	2,281
	20–24	16,146	15,433	44,295	42,338
	25–29	26,501	25,411	169,924	162,081
	30–34	15,285	14,209	143,257	135,881
	35–39	2,215	2,178	15,895	15,121
35–39	<20	831	814	681	674
	20–24	3,539	3,272	10,519	9,976
	25–29	7,033	6,665	44,878	42,921
	30–34	8,715	8,213	102,655	97,504
	35–39	4,824	4,630	66,044	63,185
40–44	<20	230	253	244	233
	20–24	1,047	974	3,247	3,137
	25–29	1,822	1,896	12,032	11,328
	30–34	3,167	3,039	31,036	29,752
	35–39	3,163	2,923	47,363	45,190
45–49	<20	134	93	106	110
	20–24	331	316	1,163	1,138
	25–29	591	528	3,959	3,855
	30–34	1,051	960	8,863	8,499
	35–39	1,234	1,145	14,889	14,472

Source: Novitsky and Kimbal (1958).

(d) Identify a model using partial associations.
(e) Based on the model identification algorithms, which model would you report?
(f) Perform a residual analysis on the selected model and at least one other. Does this modify the selection?
(g) Does the sex ratio depend on maternal age? Are these associations collapsible with respect to birth order? Summarize your findings with a short write-up.
(h) Are there any other interesting patterns in the data?

4. The data in Table 6.13 illustrate Simpson's paradox in four dimensions. For tuberculosis the death rates in Mississippi and New York are almost identical. However, they are lower in Mississippi

Table 6.13 Tuberculosis Deaths, Death Rates, and Population by State, Race, and Sex for Mississippi and New York, 1940[a]

State	Race	Sex	Population[b] (1000s)	Tuberculosis Deaths[c]	Tuberculosis Rate/100,000
Mississippi	White	Male	556	102	29.8
		Female	550	166	18.5
		Total	1,106	268	24.2
	Black	Male	528	371	70.2
		Female	549	427	77.8
		Total	1,077	798	74.1
	Total		2,184	1,066	48.8
New York	White	Male	6,408	3,409	53.2
		Female	6,471	1,618	25.0
		Total	12,880	5,023	39.0
	Black	Male	282	703	249.2
		Female	317	535	168.5
		Total	600	1,238	206.5
	Total		13,479	6,254	46.4

[a]Totals may not add because of rounding.
[b]*Source:* U.S. Bureau of the Census (1943a), *Sixteenth Census of the United States.*
[c]Estimated.
[d]*Source:* U.S. Bureau of the Census (1943b), *Vital Statistics of United States, 1900–1940.*

than New York when we consider whites and blacks separately. This pattern persists when we further subdivide the data by sex. To resolve the paradox, identify an appropriate log-linear model and determine the collapsible associations. This data set also illustrates the problem of very large numbers. Finally, since the disease is very rare, the odds ratio is a good approximation to the relative risk, so a log-linear model can be employed.

7

Logistic Regression

Logistic regression extends categorical data analysis to data sets with a binary response and one or more continuous factors. In the preceding chapters the factors and responses have each had a small number of levels; thus it was possible to array the data set as a multidimensional contingency table. Moreover, it has been assumed that the overall sample size was sufficiently large that all the cells had an expected count of at least one. In this chapter we explore analyses where one or more of the factors is ordinal and has many levels. As before, the underlying theory requires a large sample size, but often the expected counts will be less than one if the data were arrayed as a multidimensional table. The major implication is that in general the formal chi-square tests of fit are often inappropriate. Cox (1970) and others have noted that if we assume a given model, the corresponding likelihood can be used for generating tests of one or more parameters. These likelihood ratio tests are valid under quite general conditions as long as the overall sample size is large. Here large means that if we have S parameters, we require $n > 10(S + 1)$.

A second advantage of logistic regression is that in practice the analysis and interpretation are quite similar to the well-known procedures of multiple regression (Draper and Smith 1981 and Kleinbaum and Kupper 1978). Thus techniques for model selection and hypothesis testing follow a well-known pattern. The two principal differences are in computations and residual analysis. Generally, the computations are much more difficult and time consuming, especially if many independent variables are considered. Second, since the observed values are either 0 or 1, plotting residuals against predicted values is often quite uninformative. A problem analogous to this is that no good measure of explained variation such as R^2 is available.

The third reason for studying logistic regression is that it provides a convenient forum for comparing several categorical data analysis procedures. The first comparison is that of factor-response models to multiple-response models. Specifically, when all the factors are categorical, the logistic regression model is equivalent to a log-linear model containing the corresponding multifactor interactions. It has been argued that when this term is nonsignificant, more powerful tests can be obtained by deleting it (Bishop 1969). In addition, the collapsibility theorems indicate that when the multifactor term is automatically retained, the chances of simplifying the results of data analysis are reduced.

The second comparison is among techniques of estimation. Two possibilities have been considered so far: maximum likelihood and weighted least squares. It has been pointed out several times that when the sample sizes are adequate, the two are equivalent and optimal. However, with binary responses it is possible to use ordinary least squares (OLS) as a computationally efficient algorithm. As noted by many authors, including Cox (1970), when the underlying probabilities of a positive response lie between 0.2 and 0.8, the three techniques generally give similar results. Hence OLS is an interesting alternative, especially in the preliminary stages of data analysis (see also Magidson 1975).

The history of logistic regression goes back a number of years and is comparatively rich. It has been proposed as an alternative to probit analysis in bioassay (Finney 1972). Much of the initial work in this area was done by Berkson (1944, 1953, 1955). It has been recognized as an alternative to discriminant analysis when normal distribution assumptions are not appropriate (Truett et al. 1967, Halperin et al. 1971). Koch (1975) advocated it as a technique for removing interaction in additive probability models. A variety of authors present detailed discussions of the computational issues (Haberman 1978, Walker and Duncan 1967). This is only a partial list of the pertinent references; many more are found in Cox (1970) and Haberman (1978).

Section 7.1 begins by considering a single continuous factor and a small sample that will illustrate the basic equations. Next, a quantal bioassay is examined to emphasize the analogy to analysis of covariance (Section 7.2). In Section 7.3, several factors, some continuous and one categorical, are examined to illustrate the model-fitting procedure as well as one of the computational hazards. Next, logistic regression is discussed in Section 7.5 in the context of "causal inference" or path analysis, as suggested by Fienberg (1980). Finally, the technical issues of comparing log-linear models to logistic regression, alternative techniques of estimation, and transformations to eliminate interaction are examined in Section 7.6.

7.1 ONE CONTINUOUS FACTOR

Factors affecting the remission of cancer are of great interest to researchers attempting to ascertain an appropriate treatment regimen. The data in Table 7.1 show *LI* values for 27 cancer patients of whom 9 went into remission. Of the 13 patients with $LI < 0.9$, none had a remission, yet among the 10 above 1.0, 6 had a remission. One approach to ascertaining whether there is a significant association is to formulate a probability model.

Let Y_h be a binary random variable that takes the value 1 if the patient goes into remission and 0 otherwise, $h = 1, \ldots, 27$. Assume that the Y_h are independent. Let the probability of remission for the hth patient be denoted q_h, $h = 1, \ldots, 27$. The linear logistic model is formulated in terms of these probabilities:

$$P(Y_h = 1) = q_h = \frac{\exp(B_0 + B_1 x_h)}{1 + \exp(B_0 + B_1 x_h)}, \tag{7.1}$$

where x_h is the *LI* value for the hth patient and B_0 and B_1 are unknown parameters. Notice that we have implicitly assumed that q_h is constant

Table 7.1 Remission of Cancer and *LI* Values: Counts

LI (x)	Remission: No	Remission: Yes	Total
0.4	2	0	2
0.5	2	0	2
0.6	3	0	3
0.7	3	0	3
0.8	3	0	3
0.9	0	1	1
1.0	1	2	3
1.1	1	1	2
1.2	1	0	1
1.3	0	1	1
1.4	0	1	1
1.5	0	0	0
1.6	1	0	1
1.7	0	1	1
1.8	0	0	0
1.9	1	2	3
Total	18	9	27
$\bar{x}$	0.83	1.36	1.00
s	0.40	0.39	0.39

Source: Harrell (1980).

for distinct values of x_h. This somewhat complicated equation can be inverted to yield the more familiar log-odds representation:

$$l_h = \log\left(\frac{q_h}{1 - q_h}\right) = B_0 + B_1 x_h. \tag{7.2}$$

The log odds, or l_h, is also referred to as the *logit*. It is now clear why this model is called the linear logistic model: l_h is a linear function of the factor x_h.

The first problem is estimating the model parameters. The data are sparse in the sense that there are never more than three observations for any given value of x_h in Table 7.1. This means that it is not possible to estimate l_h directly. However, the likelihood function provides a way around this difficulty. Specifically, for each observation there is a *Bernoulli trial,* that is, a binomial with one trial, so

$$P(Y_h = y_h) = q_h^{y_h}(1 - q_h)^{1-y_h} \qquad h = 1, \ldots, n. \tag{7.3}$$

By assumption the Y_h are independent, so

$$\begin{aligned}
P(Y_1 = y_1, Y_2 = y_2, \ldots, Y_n = y_n) &= P(Y_1 = y_1)P(Y_2 = y_2) \cdots P(Y_n = y_n) \\
&= \prod_{h=1}^{n} q_h^{y_h}(1 - q_h)^{1-y_h} \\
&= \prod_{h=1}^{n} \left[\frac{\exp(B_0 + B_1x_h)}{1 + \exp(B_0 + B_1x_h)}\right]^{y_h} \left[\frac{1}{1 + \exp(B_0 + B_1x_h)}\right]^{1-y_h} \\
&= \frac{\Pi_{h=1}^{n} [\exp(B_0 + B_1x_h)]^{y_h}}{\Pi_{h=1}^{n} [1 + \exp(B_0 + B_1x_h)]}
\end{aligned} \tag{7.4}$$

The term on the right-hand side of (7.4) is termed the "likelihood of observing $(y_1, y_2, \ldots, y_n)$." The maximum likelihood estimates are those values of B_0 and B_1 which produce the largest value of (7.4) for the observed sample. Usually, it is easier to work with the log-likelihood:

$$\begin{aligned}
L &= \log\left\{\prod_{h=1}^{n} [\exp(B_0 + B_1x_h)]^{y_h}\right\} \\
&\quad - \log\left\{\prod_{h=1}^{n} [1 + \exp(B_0 + B_1x_h)]\right\} \\
&= B_0 \sum_{h=1}^{n} y_h + B_1 \sum_{h=1}^{n} x_h y_h - \sum_{h=1}^{n} \log[1 + \exp(B_0 + B_1x_h)]
\end{aligned} \tag{7.5}$$

The first two sums in the log-likelihood will only be affected by cases with remission, $y_h = 1$. Unfortunately, the third sum will have contributions from each distinct value of x_h, so finding the maximum of L, with respect to B_0 and B_1, is no easy task. Before tackling it, we consider a simpler case.

Let $\Sigma_{h=1}^{n} y_h = Y$, $z_h = 1$ if $x_h \geq 1$, and $\Sigma_{h=1}^{n} z_h = Z$. Now replace x_h by z_h in (7.5) and let $Y_1 = \Sigma z_h y_h$. The result is

$$L = YB_0 + Y_1 B_1 - (n - Z) \log [1 + \exp (B_0)] - Z \log [1 + \exp (B_0 + B_1)].$$

Differentiate this with respect to B_0 and B_1 and set the results equal to zero. This yields two equations in two unknowns which can be solved to yield (after some algebra)

$$\begin{aligned} B_1 &= \log [Y_1(n - Y - z + Y_1)] - \log [(Y - Y_1)(Z - Y)] \\ B_0 &= \log (Y - Y_1) - \log (n - Y - Z + Y_1). \end{aligned} \qquad (7.6)$$

Hence, if the independent variable is dichotomous, B_1 is the familiar log odds ratio and B_0 is the conditional log odds of remission. Alternatively, B_1 corresponds to the log odds ratio for remission given a unit shift in *LI*.

Returning to the original problem, it is necessary to maximize (7.5) using iterative methods such as Newton–Raphson or "iterated least squares." The former is illustrated in detail by Haberman (1978, Chap. 5). The latter has been discussed by Baker and Nelder (1978). Newton–Raphson has been implemented in BMDP and SAS. The iterated least squares procedure is available in SAS and GLIM. The resulting estimates are $b_0 = -3.777$ (SE = 1.379) and $b_1 = 2.897$ (SE = 1.187) and $L = -13.035$. Thus the predicted probabilities of remission are given by (7.1):

$$\hat{q} = \frac{\exp (-3.777 + 2.897x)}{1 + \exp (-3.777 + 2.897x)}.$$

A unit change in *LI* increases the odds of remission by $\exp (b_1) = 18.12$, but note that the range in *LI* was only observed to be 0.4 to 1.9 with a standard deviation of $s_p = 0.3931$ (see Table 7.1).

The next issue is testing the significance of B_1. There are two approaches. The first is to compare the likelihood with B_1 estimated, as above, to the likelihood with $B_1 = 0$, which can be computed by hand. First maximizing the likelihood yields the *restricted estimate* of B_0:

$$b_0^R = \log\left(\frac{Y}{n}\frac{2}{n - Y}\right) = \log\left(\frac{9}{18}\right) = \log\left(\frac{1}{2}\right) = -0.69.$$

Next we compute L^R:

$$\begin{aligned} L^R &= b_0^R \sum_{h=1}^{n} y_h - n \log\left[1 + \exp\left(b_0{}^R\right)\right] \\ &= -0.69(9) - 27 \log\left[1 + \exp\left(-0.69\right)\right] = -17.186. \end{aligned}$$

Now the likelihood ratio test is given by $2(L - L^R) = 34.37 - 26.07 = 8.30$. Since only one parameter is hypothesized to be zero, df $= 1$. It can be shown that if the null hypothesis is true, then $G^2 = 2(L - L^R)$ is chi-square. Hence we conclude that B_1 is significantly different from zero.

The second approach to hypothesis testing is to compare b_1 to its standard error. This is the Wald test that we have previously examined. The computation of the standard errors again requires the use of computer programs and will not be discussed (see Haberman 1978). In this case the Wald test is $Q = (2.897/1.187)^2 = 5.96$, which is also to be compared with a chi-square, df $= 1$, distribution. The large discrepancy between Q and G^2 is due to the small sample size.

Finally, using the standard error, confidence intervals on the odds ratio can also be constructed. For the 95 percent interval the confidence coefficient is $z = 1.96$. The corresponding interval is

$$\begin{aligned} CI &= \{\exp\left[b_1 - 1.96(\text{SE})\right], \exp\left[b_1 + 1.96(\text{SE})\right]\} \\ &= \{\exp\left[2.897 - 1.96(1.187)\right], \exp\left[2.897 + 1.96(1.187)\right]\} \\ &= (1.77, 185.57). \end{aligned}$$

This does not include 1.0, so it is consistent with the hypothesis tests. The extreme width emphasizes the lack of precision due to the small sample size.

Before turning to more complex examples it is worthwhile to consider two other approaches to analyzing these data. One possibility is to dichotomize *LI*, the other is to assume that the values are normally distributed. The first approach converts the data set into a 2×2 table. One possible cut point is at $x_h = 1$, as discussed above. This yields the first set of values in Table 7.2. Here the chi-square tests are similar to the logistic regression model values. Also, the odds ratio is 20.8, similar to the value based on b_1. Table 7.2 shows also what happens if $LI = 1.5$ is used as a cut point. Now the chi-square values are much less than the critical value of 3.84. Moreover, the odds ratio is 4.0. This is an example of how categorizing a continuous variable can produce very different results, depending on where the cut point occurs. In general, the logistic regression model is to be preferred when one or more of the factors is continuous.

The second approach is to assume that the *LI* values are normally distributed with common variance. The estimated means and standard errors are shown in Table 7.1. The corresponding t-statistic is

$$t = \frac{\bar{x}_1 - \bar{x}_0}{\sqrt{s_p^2(1/n_1 + 1/n_0)}}$$

$$= \frac{1.36 - 0.83}{\sqrt{(0.393)^2(\frac{1}{9} + \frac{1}{18})}} = 3.30.$$

Table 7.2 Cancer Remission as a Response to Dichotomized *LI*

LI Values	Cancer Remission			
	No	Yes	Total	
		Cut Point = 1		
<1.0	13	1	14	
1.0+	5	8	13	$X^2 = 8.98$
Total	18	9	27	$G^2 = 9.84$
		Cut Point = 1.5		
<1.5	16	6	22	
1.5+	2	3	5	$X^2 = 1.96$
Total	18	9	27	$G^2 = 1.86$

This is highly significant when compared to a t-distribution with df = 9 + 18 − 1 = 26. In general, when normal assumptions apply to all the factors, then t-tests or discriminant analysis is to be preferred to logistic regression. However, if one or more of the factors is categorical, logistic regression is preferred (Halperin et al. 1971).

7.2 LOGISTIC COVARIANCE ANALYSIS: QUANTAL BIOASSAY

The comparison of two (or more) groups after adjusting for some factor is one of the most common problems in data analysis. When the response and adjustment factor are continuous, the usual term is *covariance analysis*. When the response is binary, this is termed *quantal assay*. The term *bioassay* is used when the subjects are living material. An example of such a quantal bioassay is shown in Table 7.3. Here 20 mice were injected with varying dosages of a standard drug and a new or treatment drug. There are five dosages for the standard and four dosages for the treatment. Notice that in each case the dosages follow a geometric progression. The number of surviving and dead mice are shown

Table 7.3 Counts, Proportions, and Logits for Quantal Bioassay

		Counts				
Drug	Dosage	Alive (1)	Dead (2)	Proportion (3)	Logit (4)	Variance (5)
Standard	1	19	1	0.05	−2.94	1.05
	2	15	5	0.25	−1.10	0.27
	4	11	9	0.45	−0.20	0.20
	8	4	16	0.80	1.39	0.31
	16	1	19	0.95	2.94	1.05
Treatment	1	16	4	0.20	−1.39	0.31
	2	12	8	0.40	−0.41	0.21
	4	5	15	0.75	1.10	0.27
	8	2	18	0.90	2.20	0.56

Source: Koch (1975).

in columns (1) and (2). What is of interest is to compare the relative potency of the treatment after adjusting for the various dosages. This raises three questions:

1. What is an appropriate model for the proportion dying?
2. Is the dosage effect the same in the two groups?
3. Is the treatment significantly different from the standard, given that the dosage is the same?

At the most basic level the response is assumed to be binomial within each drug dosage combination. Since there are no observed zeros, the data will be used to illustrate weighted least squares estimation, although maximum likelihood estimation is just as feasible.

Let Y_{ij} denote the number of deaths at the jth dosage for the ith drug and n_{ij} the corresponding sample size ($n_{ij} = 20$ for all i and j, which need not be true). The underlying probability of death is given by

$$P(\text{death} \mid i\text{th drug}, j\text{th dosage}) = q_{ij} > 0. \tag{7.7}$$

Let d_{ij} denote the dosage for the ith drug and jth treatment. Since the dosages follow a geometric progression, a log transformation will convert this to a linear progression. Finally, one possible model for the quantal reponse is the logistic. Others are discussed in detail by Finney (1972).

The logistic model is basically the same as before, except that a categorical variable, drug, has been added:

$$\log\left(\frac{q_{ij}}{1 - q_{ij}}\right) = X_1 D_1 + X_2 D_2 + X_3 B_1 + X_4 B_2, \tag{7.8}$$

where

$$X_1 = 1 \quad \text{if } i = 1 \text{ (standard)}, \quad 0 \text{ otherwise}$$

$$X_2 = 1 \quad \text{if } i = 2 \text{ (treatment)}, \quad 0 \text{ otherwise}$$

$$X_3 = \log_2 (d_{ij}) \quad \text{if } i = 1, \quad 0 \text{ otherwise}$$

$$X_4 = \log_2 (d_{ij}) \quad \text{if } i = 2, \quad 0 \text{ otherwise.}$$

Here D_1 and D_2 are the intercepts for the two drugs, and B_1 and B_2 are the slopes of the $\log_2$ (dose) effects. To emphasize that the experiments are independent, the model has been written without an overall intercept. If there is a common dose effect (H_0: $B_1 = B_2$), the treatment effect will be the difference of the intercepts: $D_1 - D_2$.

The nine logits are shown in column (4) of Table 7.3. Since there are no observed zeros, all of the logits can be estimated. An alternative suggested by Cox (1970) is to analyze the *empirical logit*:

$$L'_{ij} = \log\left(\frac{\hat{q}_{ij} + 1/2n_{ij}}{1 - \hat{q}_{ij} + 1/2n_{ij}}\right).$$

This is slightly less biased than the logit, but in practice the difference is small.

The complete design matrix and estimated model parameters as well as fitted values are shown in Table 7.4. Here X_3 and X_4 correspond to log dose, where the logarithm is base 2 for convenience. The estimates are computed according to the formulas in Appendix 4A.1. The estimated variances of the logits are

$$\text{Var}\,(L_{ij}) = \frac{n_{ij}}{Y_{ij}(n_{ij} - Y_{ij})}. \tag{7.9}$$

By comparing the observed to fitted values obtained by inverting the fitted logits [column (1)], the usual fit statistic is obtained, $Q = 0.75$. Since there are four estimated parameters and nine populations, df = 9 − 5 = 4. Clearly, these data could have arisen from a logistic model.

The answer to the second question is obtained by comparing b_1 to b_2, which appear quite similar. The corresponding hypothesis is

$$H_0\colon\ B_1 - B_2 = [0 \quad 0 \quad 1 \quad -1]\begin{bmatrix} D_1 \\ D_2 \\ B_1 \\ B_2 \end{bmatrix} = \mathbf{CB} = 0. \tag{7.10}$$

Again the weighted least squares statistics is used, yielding $Q = 0.10$ and df = 1. From this it follows that the dosage effect is the same for the two drugs. This corresponds to the parallelism hypothesis in analysis of covariance.

Table 7.4 Fitted Values, Design Matrix, and Parameter Estimates for Quantal Assay

		Fitted Values		Design Matrix				Parameter				
Drug	Dosage	Initial (1)	Final (2)	X_1 (3)	X_2 (4)	X_3 (5)	X_4 (6)	Label (7)	Initial (8)	SE (9)	Final (10)	SE (11)
Standard	1	−2.71	−2.60	1	0	0	0	D_1	−2.71	0.58	−2.60	0.46
	2	−1.35	−1.30	1	0	1	0	D_2	−1.49	0.45	−1.57	0.40
	4	0.01	0.00	1	0	2	0	B_1	1.36	0.26	1.30	0.19
	8	1.36	1.30	1	0	3	0	B_2	1.24	0.28		
	16	2.72	2.61	1	0	4	0					
Treatment	1	−1.49	−1.58	0	1	0	0					
	2	−0.25	−0.28	0	1	0	1					
	4	0.98	1.03	0	1	0	2					
	8	2.22	2.33	0	1	0	3					

The similarity of the two dosage effects suggests that they should be combined before the drugs are compared. This is accomplished by adding X_3 and X_4 and reestimating the model. The corresponding fitted logits and parameter estimates are shown in columns (2) and (10), respectively. The fit statistic is $Q = 0.85$, df $= 9 - 3 = 6$, so this model is very good indeed. Comparing the drugs now corresponds to

$$H_0\colon\ D_1 - D_2 = [1 \quad -1 \quad 0]\begin{bmatrix} D_1 \\ D_2 \\ B \end{bmatrix} = 0.$$

The Q-statistic is 6.49, df $= 1$, so the drugs are significantly different.

The difference between the two drugs is often summarized in terms of the "relative potency" of the treatment. This is defined to be

$$P = \exp\left(\frac{D_2 - D_1}{B}\right) \tag{7.11}$$

or

$$P = \exp\left(\frac{-1.57 + 2.60}{1.30}\right) = 2.20.$$

The results can be summarized as follows:

1. A logistic model is appropriate for this quantal assay.
2. The dosage effect is the same for both drugs.
3. The treatment is twice as potent as the standard.

With a bit more computation, confidence intervals for the potency can also be obtained. For more details one should consult the Berkson references, and for a defense of the potency definition, Finney (1972). The important point is that the logistic model has now been extended to the mixed-factor case, one categorical factor and one continuous factor. Moreover, the analysis can be treated in a manner analogous to multiple regression.

7.3 MODEL SELECTION

In simple terms logistic regression is just like multiple regression, with three exceptions. First, the response is the logit of a Bernoulli random variable. Second, the computation of estimates is more cumbersome in the sense that it requires more time on a computer. Third, the parameters are interpreted in terms of effects on log odds rather than on the means of the response variables. However, the underlying similarity is close enough to allow using techniques originally developed for multiple regression. This can be best illustrated with an example.

The government annually collects data on *health maintenance organizations* (HMOs). In general terms, an HMO is an association of physicians that provides a broad range of medical care services to a voluntarily enrolled population, for a fixed fee. There are two types of HMOs: *prepaid group practice* (PGP) and *independent practitioner association* (IPA). In the PGP, physicians representing many specialties practice together in a single facility and are remunerated by salary or some form of capitation. Physicians organized in an IPA practice in their own offices and bill the association for their services according to a predetermined ''fee for service'' schedule. The IPA is a newer form of HMO and it is of considerable interest to know whether it differs from the PGP with respect to the hospitalization rates of members or subscribers.

Some of the variables collected by Interstudy in 1980 are shown in Table 7.5. These include HMO location (Region), hospital admissions per 1000 members (ADMI80), average length of stay (ALOS80), total members (ENRH80), years the organization has operated (OPER80), mean visits per member (VSTH80), and hospital days per 1000 members. The data were collected for 260 HMOs, of which 153 were PGPs and 107 were IPAs. The objective of the analysis is to determine whether there is a difference between PGPs and IPAs in their hospitalization rates. The limitations of this approach will be discussed after an appropriate model is selected.

This data set is observational and in some sense a case-control or retrospective study. Basically, the sample can be thought of as stratified according to type of plan IPA (case) or PGP (control). The variables are then measured retrospectively. This suggests logistic regression as a method of analysis. Let $Y_h = 1$ if IPA and 0 otherwise be the dichotomous response. Thus we are interested in whether the hospitalization rate is associated with the type of HMO. Moreover, it is important to ascertain which of the remaining variables affect this association.

Table 7.5 Descriptive Characteristics of 260 Health Maintenance Organizations

Variable Name	Label	Percent Missing	Mean	Statistics S.D.	Skewness	Comment
Region						
North	X_1	0	0.23	Categorical variable region reparameterized to three dummy variables		
Central	X_2	0	0.31			
South	X_3	0	0.19			
West	—	0	0.27			
ADMI80	X_4	45.4	110.2	134.3	10.6	
ALOS80	X_5	45.4	5.0	1.5	2.9	Right skewed
ENRH80	X_6	9.2	38,558.8	15,848.3	9.0	
OPER80	X_7	10.4	7.0	11.5	6.3	
VSTH80	X_8	39.6	3.2	0.7	0.4	
HDAY80	X_9	23	450.0	120.8	0.2	"$X_4 \cdot X_5$"

Source: Interstudy (1980).

Descriptive statistics are provided for each variable in Table 7.5. The variable "region" is nominal with four levels, so it is rewritten in terms of three dummy variables (X_1, X_2, X_3), comparing each region to the West. This comparison is quite arbitrary and only affects the interpretation of the results. However, it is important that these three dummy variables be examined jointly since a different parameterization will yield different estimates, and hence different individual test statistics.

The remaining variables ($X_4 - X_9$) are all continuous but have several peculiarities. First, $X_4 - X_7$ are extremely skewed, especially X_4 and X_6. This is not generally a problem for multiple regression, but the iterative methods used for maximum likelihood estimation of the logistic regression parameters may be affected. Second, X_9 is the product of X_4 and X_5. But the data reporting was often quite incomplete, so some plans reported only one of these variables, hence the remaining two could not be constructed. Finally, four variables have missing value rates that exceed 20 percent (X_4, X_5, X_8, X_9). This means that the results of any analysis need to be interpreted with great caution.

With these limitations in mind, an attempt is made to estimate the following model:

$$\log\left[\frac{P(Y=1)}{P(Y=0)}\right] = B_0 + B_1X_1 + B_2X_2 + \cdots + B_9X_9,$$

that is, the log odds of an IPA. Two things occur. First, because of the pattern of missing values, only 106 of the HMOs can be utilized. That is, the missing data rate is much worse than indicated by Table 7.5. Second, the enrollment of the plans is extremely dispersed and the computational algorithm fails to converge!

Nothing can be done about the missing data but to increase our caution about interpreting the results. The second problem can be overcome by transforming the troublesome variable. One possible transformation is to use the ranked enrollment. That is, rather than using the actual enrollment, we use the relative size of the plan. Thus the largest plan receives a value of 260 and the smallest 1. We denote the value of the ranked enrollment as RX_6.

The estimation algorithm is run on the new data set, yielding the results shown in Table 7.6. Since there are relatively few variables, a backwards elimination procedure is employed. Thus all nine variables are considered and the likelihood ratio relative to no variables is 35.47. Each variable is tested for significance and the one producing the smallest change in G^2 is deleted. In this data set it is X_4 (hospital admissions per 1000 members).

Table 7.6 Results of Stepwise Regression Analysis on 106 Health Maintenance Organizations with Complete Data

Step	Model	df	G^2	Δdf	ΔG^2	Deleted Variable
0	X_1–X_9	9	35.47			
1	X_1–X_3, X_5–X_9	8	35.47	1	0.0	ADMI80 ($X4$)
2	X_1–X_3, RX_6–X_9	7	35.46	1	0.01	ALOS80 ($X5$)
3	X_1–X_3, RX_6, X_7, X_9	6	34.74	1	0.72	VSTH80 ($X8$)
4	X_1–X_3, X_7, X_9	5	32.56	1	2.18	RENR80 ($RX6$)

Source: Harrell (1980).

All variables are again tested for significance and that producing the smallest change is dropped. This procedure continues through four steps, eliminating in order X_4, X_5, X_8, and RX_6. All remaining variables achieve statistical significance ($p < 0.05$). This suggests that the odds of an IPA are associated with region (X_1, X_2, X_3), years operational (X_7), and hospital days per 1000 members. The overall summary statistics for this model are shown at the bottom of Table 7.7. Since the reduction in $G^2 = 2.91$ with df $= 4$ it is evident that nothing significant has been eliminated.

An important point to note is the variables with the highest missing value rates have been fortuitously eliminated from the model. This means that some idea of the effect of the missing data can be obtained by reestimating the model parameters using the more complete data set.

It turns out that 197 HMOs have complete data for $X_1 - X_3$, X_7, and X_9. The two sets of estimates for $b_0 - b_3$, b_7, and b_9 are shown at

Table 7.7 Estimates for 106 Complete Observations and 197 Partial Observations

		$n = 106$			$n = 197$		
Variable		Estimate	S.E.	Q	Estimate	S.E.	Q
Intercept		−0.47	1.19	0.15	−1.00	0.66	2.33
North	(X_1)	−1.86	0.68	7.45	−1.42	0.45	10.15
Central	(X_2)	−2.11	0.73	8.27	−1.22	0.45	7.53
South	(X_3)	−3.26	1.00	10.71	−1.66	0.51	10.68
OPER80	(X_7)	−0.27	0.09	8.47	−0.07	0.03	7.40
HDAY80	(X_9)	0.01	<0.01	4.82	<0.01	<0.01	10.33

Summary Chi-square Statistics

Source	df	G^2 ($n = 106$)	G^2 ($n = 197$)
Model	5	32.56	
Deleted terms	4	2.91	NA
Total	9	35.47	NA

the top of Table 7.7. The first point is that for $n = 106$ and $n = 197$ all variables are significant (except the intercept, b_0, which is nonsignificant for both models). Second, increasing the sample size has the expected effect of reducing the standard errors of all the estimates. Third, the direction of the effects is the same for all the estimates. Thus being in any region other than the West reduces the odds of an IPA. Similarly, older plans (more years operational) are less likely to be an IPA, as was expected. Finally, IPAs are associated with increased days of hospitalization ($b_9 > 0$).

An important qualification is that all the parameter estimates in the smaller data set are larger in absolute value than in the more complete data set. Thus increasing the data set has reduced the magnitude of the associations. We can only speculate as to the impact on the estimates of having complete data for all 260 HMOs.

Two further qualifications are important, one statistical and one substantive. The statistical qualification is that the use of logistic regression was indicated by the assumption that the data were collected retrospectively. If the data were collected prospectively, hospitalization days or rates could have been treated as a response. Since these are continuous, they (or an appropriate transformation) could have been analyzed using multiple regression. In this case, HMO type would be an independent variable. For more on this approach, see Kleinbaum and Kupper (1978).

The substantive qualification may be more serious. This analysis assumes that only HMO characteristics affect the hospitalization days and rates. It is well known that the socioeconomic, demographic, and health status composition among other variables of the HMOs have a great influence on hospital utilization patterns. Hence this analysis is at best preliminary until more extensive data are available.

7.4 CONDITIONAL PROBABILITY MODELS

The analysis of categorical data with log-linear models built additive linear models for the underlying joint probability structures. For example, a model for four dichotomous factors with a sample of size n is of the form

$$\ln (m_{hijk}) = \ln (np_{hijk}) = \ln [nP(A = h,\ B = i,\ C = j,\ D = k)]$$
$$= M + A_h + B_i + C_j + D_k$$
$$+ \text{ (various interaction terms)} \tag{7.12}$$

with appropriate identifiability conditions on the parameters for estimation purposes. Recall that the Roman letters without subscripts indicate the variables for the n subjects, while those with subscripts indicate parameters. In logistic regression the focus is on the conditional probabilities of a single response given the factors

$$\ln \left(\frac{q_{hijk}}{1 - q_{hijk}}\right) = \ln \left[\frac{P(D = k|A = h,\ B = i,\ C = j)}{1 - P(D = k|A = h,\ B = i,\ C = k)}\right]$$
$$= \sum_{s=0}^{S} B_s X_s. \tag{7.13}$$

It will be shown in Section 7.5 that the two approaches are equivalent under certain conditions.

Suppose that the factors occur sequentially in the sense that for substantive reasons or by design, A occurs before B, B occurs before C, and C occurs before D. Here the phrase "occurs before" means that there is a strict temporal ordering or the two variables are measures at the same time. Then it is of interest to rewrite the joint probability in terms of a product of conditional probabilities following the rules in Section 2.2:

$$P(A = h,\ B = i,\ C = j,\ D = k)$$
$$= P(A = h)P(B = i|A = h)P(C = j|A = h,\ B = i)$$
$$P(D = k|A = h,\ B = i,\ C = j).$$

This suggests constructing joint probability models from the product of conditional probability models. If we apply log transforms this means that log-linear models of m_{hijk} can be thought of as sums of models for the log (q_{hijk}). This can be illustrated with an example.

In Chapter 6 a model for the joint association between mother's age and smoking, and infant gestation and survival was constructed.

The temporal ordering of these variables was not considered except on a limited basis when the results were interpreted. However, more insight can be gained by recognizing that age and smoking occur before gestation, which occurs before infant survival. As for age and smoking, they are usually measured at the same point in time, although it could be argued that age occurs before smoking. This suggests examining four probabilities:

(1)	$P(A = h)$	probability of selecting a mother age $h = 1, 2$
(2)	$P(S = i \mid A = h)$	probability that a mother age h is in smoking category i
(3)	$P(G = j \mid A = h, S = i)$	probability that a mother age h in smoking category i has an infant in gestation category j
(4)	$P(D = k \mid A = h, S = i, G = j)$	probability that an infant survives (or dies) given a mother age h, smoking category i, and gestational age j.

(7.14)

Since each of the responses are dichotomous, logistic regression seems appropriate.

The logistic models are

$$
\begin{aligned}
&(2)\quad \ln\left(\frac{q_{Sh}}{1 - q_{Sh}}\right) = M_S + A_{Sh} \\
&(3)\quad \ln\left(\frac{q_{Ghi}}{1 - q_{Ghi}}\right) = M_G + A_{Gh} + S_{Gi} \\
&(4)\quad \ln\left(\frac{q_{Dhij}}{1 - q_{Dhij}}\right) = M_D + A_{Dh} + S_{Di} + G_{Dj}.
\end{aligned}
\tag{7.15}
$$

One could also construct a model for $P(A = h)$.

Using the procedures discussed in Sections 7.1–7.3, maximum likelihood estimates of the model parameters and the corresponding tests of fit are obtained. These are shown in Table 7.8. The three

Table 7.8 Logistic Model Parameter Estimates, Standard Errors, and Wald Test for Three Conditional Probability Models

	Response		
Factors Estimate (S.E.)	Not Smoking (S)	Prematurity (G)	Infant Death (D)
Intercept	2.35 (0.05)	−2.22 (0.07)	−3.02 (0.14)
Q	2236.64	946.19	488.61
Age (A)	−0.20 (0.05)	−0.12 (0.04)	−0.23 (0.09)
Q	16.56	7.40	6.72
Not smoking (S)	—	−0.06 (0.07)	−0.21 (0.13)
Q		0.76	2.60
Prematurity (G)	—	—	1.66 (0.09)
Q			321.40
Error (df)	0.00 (0)	0.15 (1)	1.48 (4)

responses are shown across the top of the table and the factors in the stub. Note that model 2 for smoking is a saturated model, so the fit statistic is identically zero. The remaining likelihood ratio test statistics are quite small, so we can safely examine the parameter estimates.

The factor "age" is present in each model and the corresponding estimates are all significant and negative; thus younger mothers are less likely not to smoke, to have premature infants, and to have an infant death. Similarly, the estimates for not smoking are negative but nonsignificant for prematurity and infant death. Hence smoking, in this sample, is a relatively unimportant factor. Finally, prematurity has a strong positive association with infant death, as is well known.

These results can also be represented graphically in a manner analogous to path analysis. A path diagram is shown in Figure 7.1. Here the temporal ordering is implied by the arrows on the connecting segments. Moreover, each response is conditioned on all variables to its left. Finally, a global test of the three models is obtained by summing the three conditional test statistics: $G^2 = 0.0 + 0.15 + 1.48 = 1.63$, df $= 0 + 1 + 4 = 5$. More discussion of the interpretation of path models for categorical data is found in Goodman (1978, pp. 173–229 and 423–436).

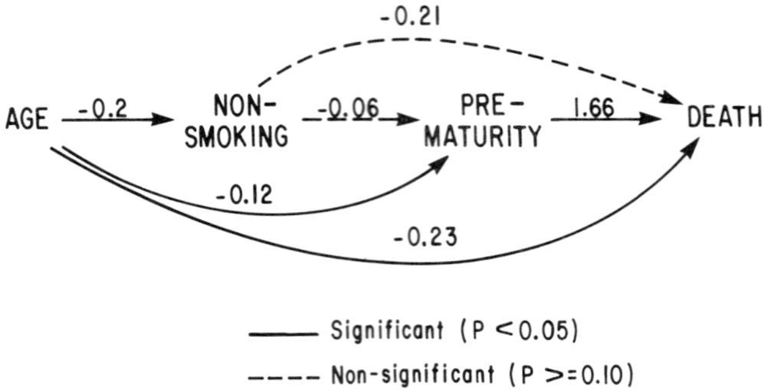

Figure 7.1 Path diagram for models of mother's smoking and age and infant's prematurity and death.

7.5 COMPARISONS USING LOGISTIC REGRESSION

Sections 7.1–7.4 have presented methods for using and interpreting logistic regression models. In this section its relationship to the log-linear model is demonstrated. Also, a logistic model is used to illustrate the similarity between maximum likelihood estimation and weighted least squares estimates. Finally, the use of the logistic model to eliminate interaction on the additive scale is illustrated. Thus the logistic model can be thought of as simply another application of an underlying strategy for categorical data analysis.

First we prove the equivalence of logistic models with categorical variables to log-linear models with a single binary response. Suppose that we have a sample of n independent and identically distributed observations each with a single binary response, Y. Moreover, each observation has a set of factors represented by a set of J dummy variables. This is another way of saying that the response variable Y is to be modeled as a function of an appropriate set of parameters. As usual, we assume that

$$P(Y = i, X_1 = x_1, \ldots, X_J = x_J) = p_{i,1,\ldots,J} > 0 \tag{7.16}$$

where $i = 1, 2$ and $x_1, \ldots, x_J$ correspond to the appropriate dummy values for the categorical variables defined by Y and the Xs. As usual, the expected counts are denoted $m_{i,1,2,\ldots,J}$. Then the log-linear model can be written

$$\ln (m_{i,1,2,\ldots,J}) = M + A_i + \sum_j C_j + \sum_j (AC)_{ij}. \tag{7.17}$$

Here, by appropriate reparameterization, the C_j ($j = 1, \ldots, J$) correspond to the main effects and interactions among the X_j ($j = 1, \ldots, J$), and the $(AC)_{ij}$ are interactions between the response and the factors. Since Y is dichotomous, $A_1 + A_2 = 0$ and $(AC)_{1j} + (AC)_{2}$ $= 0$ can be used as identifiability conditions.

The objective is to convert the log-linear model into a logit model First consider the log ratio of the expected counts:

$$\begin{aligned}
\ln\left(\frac{m_{1,1,2,\ldots,J}}{m_{2,1,2,\ldots,J}}\right) &= \ln\left[\frac{(m_{1,1,2,\ldots,J}/m_{+,1,2,\ldots,J})}{(m_{2,1,2,\ldots,J}/m_{+,1,2,\ldots,J})}\right] \\
&= \ln\left[\frac{(np_{1,1,2,\ldots,J}/np_{+,1,2,\ldots,J})}{(np_{2,1,2,\ldots,J}/np_{+,1,2,\ldots,J})}\right] \\
&= \ln\left[\frac{P(Y = 1|X_1 = x_1, \ldots, X_J = x_J)}{P(Y = 2|X_1 = x_1, \ldots, X_J = x_J)}\right] \\
&= \ln\left(\frac{q_{1\ldots J}}{1 - q_{1\ldots J}}\right)
\end{aligned}$$

where $q_{1\ldots J} = P(Y = 1|X_1 = x_1, \ldots, X_J = x_J)$. Simply put, the log ratio of the expected counts is identical to the logistic response function. Next, replace the log ratio of the expected counts by the corresponding log-linear model (7.17) employing the identifiability conditions: $A_2 = -A_1$ and $(AC)_{2j} = (AC)_{1j}$:

$$\begin{aligned}
\ln\left(\frac{m_{1,1,2,\ldots,J}}{m_{2,1,2,\ldots,J}}\right) &= \ln (m_{1,1,2,\ldots,J}) - \ln (m_{2,1,2,\ldots,J}) \\
&= \left[m + A_1 + \sum_j C_j + \sum_j (AC)_{ij}\right] \\
&\quad - \left[m + (-A_1) + \sum_j C_j + \sum_j (-AC)_{1j}\right] \\
&= 2A_1 + 2\sum_j (AC_{1j}) = \ln\left(\frac{q_{1\ldots j}}{1 - q_{1\ldots j}}\right).
\end{aligned}$$

But this is the usual logistic model:

$$\ln\left(\frac{q_{1\ldots J}}{1 - q_{1\ldots J}}\right) = B_0 + \sum_{j=1}^{J} B_j x_j \quad \text{where } B_0 = 2A_1 \text{ and } 2(AC)_{1j} = B_j x_j.$$

Notice that the identifiability conditions in the B_j are contained in the X_j. Hence any logistic model with only categorical factors is identical to a log-linear model which contains all of the interactions among the factors, and conversely. As a corollary, using the logistic model precludes estimation of the associations among the factors of the logit model. The data from the previous section can illustrate this point.

On the left of Table 7.9 are the maximum likelihood estimates and Wald tests for the log-linear model denoted [*SAG, OS, OA, OG*]. On the right are the corresponding logistic model parameters and Wald tests from Table 7.8. The ratio of the estimates is exactly 2. Also, the change in G^2 (center of the table) is identical for both approaches. The discrepancy in Wald tests is due to the more approximate method of estimating standard errors employed by the log-linear model program.

From the proof and illustration it follows that under appropriate conditions logistic models are log-linear models. The choice depends on three questions:

Table 7.9 Comparison of Log-Linear and Logistic Model: Estimates and Test Statistics

Log-Linear Model[a]				Logistic Model[b]		
Label	Parameter Estimate	Wald Test	ΔG^2	Label	Parameter Estimate	Wald Test
O_1	−1.512	—	—	B_0	−3.024	—
OA_1	−0.117	6.72	6.53	B_1	−0.234	6.72
OS_1	−0.106	2.59	2.40	B_2	−0.211	2.60
OG_1	0.827	321.38	337.68	B_3	1.655	321.40
Fit (df)	1.48 (4)				1.48 (4)	

[a]Computed using BMDP4F (Dixon 1981).
[b]Computed using PROC FUNCAT: ML option (SAS 1982b).

1. Is there a single dichotomous response?
2. Are the factors all categorical?
3. Are associations among the factors of interest?

With three yes-or-no questions there are eight possible situations. However, if there is no clear-cut dichotomous response, log-linear models are usually the preferred approach. If associations among the factors are of interest, this is similar to the no-single-response situation, so again log-linear models are preferred. Finally, if one or more factors are truly continuous, logistic models are preferred.

The analyst should make a careful and defensible decision in each situation. The important distinction is the use of a logistic model may result in some loss of power (Bishop 1969) and may preclude collapsing over one or more variables. On the other hand, categorizing a continuous variable introduces an element of arbitrariness into the analysis (see Section 7.1). The best guide is, as usual, a thorough understanding of the study design and objectives.

A second comparison of interest is logistic models estimated by maximum likelihood or by weighted least squares (w.l.s.). The most critical difference is that weighted least squares can only be employed when the data can be arrayed as a multiway table with few, if any, zero cell counts. Moreover, the zero counts must be replaced by some arbitrary quantity, such as 0.5. This means that with continuous factors m.l.e.'s must be employed. The advantage of the w.l.s. estimators is that they are generally computationally simple and they easily permit changing the response function from the logit to the log relative risk or simple difference.

The similarity between w.l.s. estimators and m.l.e.'s can be illustrated with Bartlett's (1935) data shown in Table 4.1. The resulting estimates and test statistics for a logistic model are shown in Table 7.10. As can be seen for samples of this size, the estimates and test statistics are nearly identical. For situations where the estimates and test statistics differ substantially, one must question whether the sample sizes are large enough to justify either procedure.

A third point of interest about logistic regression is that the logit transformation of the response proportion can be used to eliminate interaction. In the absence of study design and other considerations, this can be a useful method of obtaining simple models for summarizing a data set. As an example, consider the hypothetical data in Table

Table 7.10 Logistic Model for Plum Root Stalk Data (Table 4.1)

Label	Maximum Likelihood Estimate	Likelihood Ratio Test	Weighted Least Squares Estimate	Weighted Least Squares Q
M	−0.509	—	−0.506	47.66
P	0.714	105.18	0.711	92.62
L	0.509	53.44	0.506	47.66
Fit	—	2.29	—	2.26

Source: PROC FUNCAT (SAS 1982b).

7.11. For purposes of illustration the proportions are assumed known, so hypothesis testing and estimation are not relevant.

In the first population the proportion responding "yes" varies from 0.04 to 0.90, depending on the age and sex of the subpopulation. Computing the average odds of a young person saying yes compared to an old person saying yes, we obtain

$$\text{OR(young)} = \left[\left(\frac{0.04/0.96}{0.20/0.80}\right)\left(\frac{0.60/0.40}{0.90/0.10}\right)\right]^{1/2}$$

$$= [(0.167)(0.167)]^{1/2} = 0.167.$$

Similarly, the average odds ratio for a male is 0.028. Finally, the interaction is 1.00, indicating that the odds ratio for the young saying yes is the same for males and females, and conversely.

Suppose that for a different population the responses are as in the right side of Table 7.11. Now the odds ratio for young among males is $[(0.10)(0.80)]/[(0.20)(0.90)] = 0.444$ but among females is $[(0.60)(0.30)/(0.70)(0.40)] = 0.643$. These are different from each other, so we say that there is interaction on the odds ratio or logit scale.

Next, the same associations are computed assuming a simple difference model. The results are shown in the lower half of the table. The interaction term for the logit population is 0.07, indicating that the

Table 7.11 Hypothetical Data for Logit and Additive Models with Interaction

Factors		Logit Population		Additive Population	
Sex	Age Response:	Yes	No	Yes	No
Male young		0.04	0.96	0.10	0.90
Male old		0.20	0.80	0.20	0.80
Female young		0.60	0.40	0.60	0.40
Female old		0.90	0.10	0.70	0.30
Associations		*Logit Model*			
Odds young		0.167	0.535		
Odds male		0.028	0.089		
Interaction		1.000	0.831		
		Additive Model			
Difference young		−0.23	−0.10		
Difference old		−0.63	−0.50		
Interaction		0.07	0.00		

Source: Koch (1977).

first factor effect depends on the second factor. Hence there is interaction. But the interaction vanishes when the appropriate model is used.

It should be obvious that one should apply logit models to logit populations and additive models to additive populations. Unfortunately, with only samples at one's disposal, the underlying population structure may not be obvious. Statistical tests can provide some indication, but since the tests are premised on a given population structure they can never be definitive. Understanding the underlying phenomenon and study design is usually the best guide to selecting a model. Alternatively, the consensus is to select the model that requires the fewest parameters. As a corollary to this, whenever an interaction is declared to be statistically significant, it is also important to indicate on what scale and what alternatives were examined.

7.6 ANALYSIS OF TWICE-CLASSIFIED DATA*

The discussion in Section 7.5 of logistic regression has assumed a set of S independent binomial samples. This is the most common application of logistic regression. However, the technique can also be employed where responses are measured at two points in time. It is often of great interest to know whether the response distributions are constant over time and whether they exhibit a "consistent structure." These questions come under the general heading of "symmetric contingency tables."

Epidemiologists have used symmetric models for the analysis of matched-pair case-control studies (Holford et al. 1978). Sociologists have employed them for studies of social mobility (e.g., Goodman 1978, Bishop et al. 1975, Haberman, 1978, 1979). These models are known as paired-comparison models in psychology and consumer preference studies (e.g., Fienberg and Larntz 1976, Imrey et al. 1976). Finally, the relationship between symmetric models and repeated measures designs is discussed at length by Koch et al. (1977). This is only a partial list of the extensive literature on this class of models.

Unfortunately, these models are often quite difficult to interpret substantively. By examining specific examples, experience with the construction and interpretation can be gained. The example illustrates that these models can be constructed by combining a log-linear model formulation with logistic regression. This observation will also facilitate interpreting the models.

7.6.1 A Period Symmetry Model

The Connecticut Blood Pressure Survey (D'Atri et al. 1979) collected data on the treatment of hypertension at the state level from approximately 4600 individuals. The data collection included actual blood pressure measurements. A 10 percent resurvey of households was also conducted to assist in the evaluation of the survey reliability and validity. A portion of the resurvey data is shown in Table 7.12. The issue of reliability of the data can be addressed through the use of the well-known kappa statistic (Fleiss 1981). In this case it had value indicating a fair degree of reliability ($k = 0.51$). A more subtle issue concerns

*This section is based on a paper by Freeman (1981).

Table 7.12 Distribution of Initial and Follow-up Blood Pressure According To Hypertension Status: WHO Definitions

Blood Pressure	Follow-up Survey Normal	Borderline	Elevated	Total
Initial survey				
Normal	105	9	3	117
Borderline	10	12	1	23
Elevated	3	2	7	12
Total	118	23	12	152

the validity of the classification scheme. Naively, two issues can be evaluated with these data:

1. Is the classification scheme consistent?
2. Is there a systematic bias in the follow-up?

The first question focuses on whether the odds of any given classification is constant relative to other classifications. This corresponds precisely to the original hypothesis of "quasi-symmetry" originally formulated by Caussinus (1966). If this hypothesis is false, the data are "inconsistent" with the symmetry model and further analysis may be stymied. The second question focuses on whether there has been a systematic shift in the marginal distributions. That is, are the margins symmetric to each other? The latter hypothesis corresponds to what many authors refer to as "marginal homogeneity." A different usage is employed since homogeneity is reserved for the comparison of populations. Thus the components of symmetry (quasi-symmetry and marginal symmetry) address the two issues of validity previously noted.

The two components of symmetry are equivalent to the hypothesis of symmetry (Caussinus 1966). Most authors recommend fitting the models of symmetry and quasi-symmetry and then using the difference as a conditional test of marginal symmetry. It is straightforward to show the direct tests are asymptotically uncorrelated. Usually, the quasi-symmetry model is appropriate and one can proceed with the analysis of conditional marginal symmetry or bias.

The fitting of symmetry models is straightforward. We begin with the usual log-linear model and some associated identifiability conditions. Let m_{ij} be the expected count at blood pressure levels i and j at times 1 and 2, respectively, where $i = 1, 2, 3$ and $j = 1, 2, 3$. The model is given by $\ln (m_{ij}) = M + A_i + B_j = AB_{ij}$, $A_1 = B_1 = AB_{1j} = AB_{i1} = 0$. The hypotheses of symmetry are given by

$$H_S: \text{ total symmetry: } AB_{ij} = AB_{ji} \text{ and } A_i = B_i \quad \text{for all } i \neq j \tag{7.18}$$

$$H_Q: \text{ quasi-symmetry: } AB_{ij} = AB_{ji} \quad \text{for all } i \neq j \tag{7.19}$$

H_H: marginal symmetry given quasi-symmetry:

$$A_i = B_i | AB_{ij} = AB_{ji} \quad \text{for all } i \neq j \tag{7.20}$$

Since there are only two dimensions the problem is easily converted to a logistic regression model:

$$l_{ij} = \ln\left(\frac{m_{ij}}{m_{ji}}\right) = (B_j - A_j) - (B_i - A_i) + (AB_{ij} - AB_{ji}). \tag{7.21}$$

If the model of quasi-symmetry holds (HQ is true), the third parenthetical term on the right-hand side is zero. Under the identifiability condition (1) there are $(r - 1)(r - 2)/2$ degrees of freedom for testing the fit of this model or one degree freedom in the blood pressure example.

The l_{ij} are the natural logarithms of the symmetric table entries, which may be referred to as symmetric logits. It is easy to show they are asymptotically uncorrelated. If the l_{ij} are all zero, total symmetry is obtained. This may be tested directly or by considering the A and B terms. Specifically, it is evident that under quasi-symmetry and the identifiability conditions the l_{ij} can be written as simple linear combinations of $B_j - A_j$ for $j = 2, \ldots, r$. Thus, in constructing the designs for the symmetric logits, it is sufficient to consider a simple parameterization of the indices of the l_{ij}:

$$\begin{bmatrix} l_{12} \\ l_{13} \\ l_{23} \end{bmatrix} = \left[\begin{bmatrix} 1 & 0 \\ 0 & 1 \\ 0 & 1 \end{bmatrix} - \begin{bmatrix} 0 & 0 \\ 0 & 0 \\ 1 & 0 \end{bmatrix}\right] \begin{bmatrix} B_2 - A_2 \\ B_3 - A_3 \end{bmatrix} = \begin{bmatrix} 1 & 0 \\ 0 & 1 \\ -1 & 1 \end{bmatrix} \begin{bmatrix} v_2 \\ v_3 \end{bmatrix}$$

$$\mathbf{l} = (\mathbf{X}_1 - \mathbf{X}_2)\mathbf{v},$$

Table 7.13 Analysis of Symmetry of Blood Pressures

Source	df	Q	p-Value
Symmetry	3	0.37	0.95
Homogeneity	2	0.10	0.95
Quasi-symmetry	1	0.27	0.60

where $\mathbf{X}_1$ contains the indicator variables for A, $\mathbf{X}_2$ those for B, and $\mathbf{v}$ contains the parameters of marginal symmetry. Estimating this model for the data in Table 7.12 using weighted least squares, we can test the appropriate symmetry hypotheses. The statistics are in Table 7.13. As can be seen, there is insufficient evidence to reject any of the hypotheses of symmetry. Therefore, it may be inferred that at this level of classification, there is no bias in measurement and the classification system is consistent.

7.6.2 Comparison of Symmetry Across Populations

The example in Section 7.6.1 is based on a methodological problem, but the underlying model can be used to formalize substantive problems as well. The relationship between occupational mobility and migration is one that has long puzzled social scientists. Part of the problem is that data relating to these two variables are rarely collected and less frequently reported. The data shown in Table 7.14 compare the occupational category of fathers and sons for sons who remained in a community and those who migrated. The researchers concluded that "this table does not, however, sufficiently display the effects of migration" (Scudder and Anderson 1954, p. 330). As we shall see, the proper method analysis will highlight these effects. The second reason for considering these data is that they illustrate the comparison of symmetry models across populations.

The analysis proceeds by fitting models of quasi-symmetry separately for the two populations and then testing the hypothesis that the marginal symmetry parameters are equal for residents and migrants In addition, it is of intrinsic interest to compare the occupational distribu-

Table 7.14 Distribution of Fathers and Sons by Occupational Group and Migration Status[a]

Father's Occupation	Son's Occupation: Professional	Business[a]	Clerical	Skilled	Other[b]	Total
			Resident			
Professional	16	15	0	1	0	32
Business	7	85	37	27	12	168
Clerical	0	10	23	18	17	68
Skilled	0	9	12	72	36	129
Other	0	9	9	25	51	94
Total	23	128	81	143	116	491
			Migrant			
Professional	12	9	1	1	0	23
Business	21	66	23	18	6	134
Clerical	3	22	13	15	3	56
Skilled	3	27	12	51	19	112
Other	1	7	9	9	12	38
Total	40	131	58	94	40	363

[a]Counts are reconstructed from the original presentation.
[b]These categories combine several of the original paper.
Source: Scudder and Anderson (1954).

tions of the fathers and sons within the two populations. Clearly, the hypothesis that the occupational distributions are independent across generations is of no interest. We begin our analysis by formalizing the model.

The expected cell count is denoted m_{hij} and the log-linear model is given by

$$\ln(m_{hij}) = M_h + F_{hi} + S_{hj} + FS_{hij}$$

$$F_{h1} = S_{h1} = \mathrm{FS}_{h1j} = FS_{hi1} = 0 \qquad \text{for all } h,\ i,\ j. \qquad (7.22)$$

Here h indexes the migration status, i indexes the father's occupational category, and j indexes the son's occupational category. As before, the model of quasi-symmetry for the two populations is constructed from the symmetric logits:

$$\begin{aligned} l_{hij} &= \ln\left(\frac{m_{hij}}{m_{hji}}\right) = (F_{hi} - S_{hi}) - (F_{hj} - S_{hj}) \\ &= v_{hi} - v_{hj}; \qquad h = 1, 2; \quad i < j = 2, \ldots, 5. \end{aligned} \tag{7.23}$$

The test of the quasi-symmetry model can be summed over populations and attention focused on the v_{hi}, of which there are $2(r - 1) = 8$, 4 for residents and 4 for migrants. If the model fits the data, the principal hypothesis is given by

$$H_H\colon \quad v_{1i} = v_{2i} \qquad i = 2, \ldots, 5.$$

This may be termed the *hypothesis of homogeneity*. A secondary hypothesis concerns the marginal symmetry of the residents and migrants:

$$H_R\colon \quad v_{1i} = 0 \qquad i = 2, \ldots, 5$$

$$H_H\colon \quad v_{2i} = 0 \qquad i = 2, \ldots, 5.$$

The computations are accomplished by putting the symmetric entries into a logistic regression program with the appropriate design matrix:

$$\begin{bmatrix} l_{h12} \\ l_{h13} \\ l_{h14} \\ l_{h15} \\ l_{h23} \\ l_{h24} \\ l_{h25} \\ l_{h34} \\ l_{h35} \\ l_{h45} \end{bmatrix} = \left[\begin{bmatrix} 1 & 0 & 0 & 0 \\ 0 & 1 & 0 & 0 \\ 0 & 0 & 1 & 0 \\ 0 & 0 & 0 & 1 \\ 0 & 1 & 0 & 0 \\ 0 & 0 & 1 & 0 \\ 0 & 0 & 0 & 1 \\ 0 & 0 & 1 & 0 \\ 0 & 0 & 0 & 1 \\ 0 & 0 & 0 & 1 \end{bmatrix} - \begin{bmatrix} 0 & 0 & 0 & 0 \\ 0 & 0 & 0 & 0 \\ 0 & 0 & 0 & 0 \\ 0 & 0 & 0 & 0 \\ 1 & 0 & 0 & 0 \\ 1 & 0 & 0 & 0 \\ 1 & 0 & 0 & 0 \\ 0 & 1 & 0 & 0 \\ 0 & 1 & 0 & 0 \\ 0 & 0 & 1 & 0 \end{bmatrix} \right] \begin{bmatrix} F_{h2} - S_{h2} \\ F_{h3} - S_{h3} \\ F_{h4} - S_{h4} \\ F_{h5} - S_{h5} \end{bmatrix}$$

$$\mathbf{l}_h = \begin{bmatrix} 1 & 0 & 0 & 0 \\ 0 & 1 & 0 & 0 \\ 0 & 0 & 1 & 0 \\ 0 & 0 & 0 & 1 \\ -1 & 1 & 0 & 0 \\ -1 & 0 & 1 & 0 \\ -1 & 0 & 0 & 1 \\ 0 & -1 & 1 & 0 \\ 0 & -1 & 0 & 1 \\ 0 & 0 & -1 & 1 \end{bmatrix} \begin{bmatrix} v_{h2} \\ v_{h3} \\ v_{h4} \\ v_{h5} \end{bmatrix} = (\mathbf{X}_2 - \mathbf{X}_1)\mathbf{v}_h.$$

The complete design is given by

$$\mathbf{l} = \begin{bmatrix} \mathbf{l}_1 \\ \mathbf{l}_2 \end{bmatrix} = \begin{bmatrix} [\mathbf{X}_2 - \mathbf{X}_1] & \mathbf{I}_2 \end{bmatrix} \begin{bmatrix} \mathbf{V}_1 \\ \mathbf{V}_2 \end{bmatrix}.$$

Since a weighted least squares procedure was used, the observed zeros were replaced by 0.5.

The analysis of sources of variation are shown in Table 7.15a. The p-value for the combined test of quasi-symmetry is 0.25, so the fit of the model is reasonable. The hypothesis of homogeneity yields $Q = 22.94$, df $= 4$, which indicates a significant difference between migrants and residents in terms of occupational categories when fathers and sons are compared. For resident sons the hypothesis of marginal symmetry is clearly false. For migrant sons the evidence against the hypothesis is somewhat weaker. Given the rejection of the marginal symmetry hypothesis, we may consider the parameter estimate and the associated asymptotic standard errors (ASEs). Given our parameterization ($F_{hi} - S_{hi} = v_{hi}$) it is evident that the resident sons are more likely than their fathers to be classified as clerical, skilled, or other. The reverse tendency is true for migrant sons. In general, this confirms the implications of the original presentation but has the advantages of

1. Working from the initial data presentation
2. Formalizing the underlying model

Finally, by example, the methods for comparing symmetry models across populations are a straightforward extension of the usual linear models analysis.

Table 7.15 Analysis of Symmetry in Occupation Distribution of Migrant and Resident Sons

a. Analysis of Variation

Source	df	Q	p-Value
Model (marginal symmetry)	8	30.41	0.00
Residents	4	21.56	0.00
Migrants	4	8.85	0.07
Residents vs. migrants	4	22.94	0.00
Quasi-symmetry	12	14.92	0.25
Total symmetry	20	45.33	0.00

b. Parameter Estimates

	Residents		Migrants	
Label	Estimate	ASE	Estimate	ASE
Business	0.55	0.42	−0.85	0.35
Clerical	1.35	0.48	−0.90	0.11
Skilled	1.48	0.48	−1.19	0.40
Other	1.66	0.49	−0.90	0.46

7.6.3 One Population—Two Variables

The third example illustrates the analysis of two variables, each of which is measured at two points in time. These data are discussed in detail by Wexler (1979) and are derived from the Mid-Town Manhattan Study (Srole et al. 1977). Here a cohort of 683 persons were evaluated as to their physical and mental health status in 1954 and again in 1974. Physical health, in these data, means the presence or absence of one or more chronic diseases. Mental health was rated as an individual being well, mildly impaired, or moderately to severely impaired. The principal hypothesis of interest concerned whether a change in an individual's physical health status was associated with a change in that individual's

$$
\mathbf{l} = \begin{bmatrix} l_{12} \\ l_{13} \\ l_{14} \\ l_{15} \\ l_{16} \\ l_{23} \\ l_{24} \\ l_{25} \\ l_{26} \\ l_{34} \\ l_{35} \\ l_{36} \\ l_{45} \\ l_{46} \\ l_{56} \end{bmatrix} = \begin{bmatrix} l_{1112} \\ l_{1113} \\ l_{1211} \\ l_{1212} \\ l_{1213} \\ l_{1123} \\ l_{1221} \\ l_{1222} \\ l_{1223} \\ l_{1231} \\ l_{1232} \\ l_{1233} \\ l_{2212} \\ l_{2213} \\ l_{2223} \end{bmatrix} = \left[\begin{bmatrix} 0 & 1 & 0 & 0 & 0 \\ 0 & 0 & 1 & 0 & 0 \\ 1 & 0 & 0 & 0 & 0 \\ 1 & 1 & 0 & 1 & 0 \\ 1 & 0 & 1 & 0 & 1 \\ 0 & 0 & 1 & 0 & 0 \\ 1 & 0 & 0 & 0 & 0 \\ 1 & 1 & 0 & 1 & 0 \\ 1 & 0 & 1 & 0 & 1 \\ 1 & 0 & 0 & 0 & 0 \\ 1 & 1 & 0 & 1 & 0 \\ 1 & 0 & 1 & 0 & 1 \\ 1 & 1 & 0 & 1 & 0 \\ 1 & 0 & 1 & 0 & 1 \\ 1 & 0 & 1 & 0 & 1 \end{bmatrix} - \begin{bmatrix} 0 & 0 & 0 & 0 & 0 \\ 0 & 0 & 0 & 0 & 0 \\ 0 & 0 & 0 & 0 & 0 \\ 0 & 0 & 0 & 0 & 0 \\ 0 & 0 & 0 & 0 & 0 \\ 0 & 1 & 0 & 0 & 0 \\ 0 & 1 & 0 & 0 & 0 \\ 0 & 1 & 0 & 0 & 0 \\ 0 & 1 & 0 & 0 & 0 \\ 0 & 0 & 1 & 0 & 0 \\ 0 & 0 & 1 & 0 & 0 \\ 0 & 0 & 1 & 0 & 0 \\ 1 & 0 & 0 & 0 & 0 \\ 1 & 0 & 0 & 0 & 0 \\ 1 & 1 & 0 & 1 & 0 \end{bmatrix} \right] \begin{bmatrix} \mathbf{a}_1 \\ \mathbf{a}_2 \\ \mathbf{a}_3 \\ \mathbf{a}_4 \\ \mathbf{a}_5 \end{bmatrix} = \begin{bmatrix} 0 & 1 & 0 & 0 & 0 \\ 0 & 0 & 1 & 0 & 0 \\ 1 & 0 & 0 & 0 & 0 \\ 1 & 1 & 0 & 1 & 0 \\ 1 & 0 & 1 & 0 & 1 \\ 0 & -1 & 1 & 0 & 0 \\ 1 & -1 & 0 & 0 & 0 \\ 1 & 0 & 0 & 1 & 0 \\ 1 & -1 & 1 & 0 & 1 \\ 1 & 0 & -1 & 0 & 0 \\ 1 & 1 & -1 & 1 & 0 \\ 1 & 0 & 0 & 0 & 1 \\ 0 & 1 & 0 & 1 & 0 \\ 0 & 0 & 1 & 0 & 1 \\ 0 & -1 & 1 & -1 & 1 \end{bmatrix} \mathbf{a}
$$

Figure 7.2 Model of quasi-symmetry for two variables twice classified.

Table 7.16 Physical and Mental Health at Two Points in Time

		1974					
Physical:		Well			Ill		
	Mental:	Well	Mild	Moderate	Well	Mild	Moderate
1954							
Well	Well	58	37	7	8	17	4
	Mild	55	83	38	11	23	19
	Moderate	10	47	48	3	13	30
Ill	Well	0	2	1	9	6	1
	Mild	6	9	3	5	22	17
	Moderate	6	5	15	1	22	42

Source: Wexler (1979).

mental health status. This hypothesis can be tested by the hypothesis of marginal homogeneity conditional on the model of quasi-symmetry.

The model of quasi-symmetry may be tested by forming the logits of the symmetric table entries and fitting an appropriate linear model. The model will correspond to the difference in a two-factor analysis-of-variance design for the upper triangle cell counts (Figure 7.2).

Here a_1 corresponds to homogeneity of the physical health margin, a_2 and a_3 to homogeneity of mental health margins, and a_4 and a_5 to

Table 7.17 Analysis of Symmetry for Mid-Town Manhattan Study

Source	df	Q	p-Value
Total symmetry	15	43.12	0.00
Physical health	1	4.69	0.03
Mental health	2	5.31	0.07
Association	2	1.00	0.61
Quasi-symmetry	10	6.69	0.75

the association of the two. Again, the model was fitted using a weighted least squares logistic regression program where 0.5 was added to the zero cell.

The results of the analysis are shown in Table 7.17. As can be seen, the model of quasi-symmetry fits quite well, $Q = 6.69$, df = 10, and the data are asymmetric, $Q = 43.12$, df = 15, hence there is marginal heterogeneity. There is no evidence of an association between the changes in the two indices. Physical health shows a significant change over time and mental health shows a nearly significant change over time. A more detailed analysis of these data reveals that physical health declines and mental health improves. The former is expected, but the latter may be influenced by potential sampling biases.

8

Small Samples—Complex Samples

All of the preceding methods for categorical data analysis have relied on two assumptions: independence in the sampling and large sample sizes. In many situations the latter assumption may be questionable. There is a class of procedures termed *exact tests* which can be usefully employed here. The basic idea is to enumerate all possible outcomes of a given study under the null hypothesis and then determine how likely the observed result is. Thus the probability of the observed sample is, in a certain sense, exactly determined.

These procedures have several important drawbacks. First, since the possible outcomes are enumerated, the tests become computationally formidable as sample size increases. Fortunately, the tests tend to behave like the usual chi-square tests in this situation.

The second drawback is that the term "exact tests" suggests that these tests are in some way preferable to large-sample or "approximate tests." Statistical theory indicates that both approaches are in the same class of large-sample tests and are optimal. For small samples the exact tests yield discrete probability distributions, so they rarely achieve the prespecified significance levels. Hence either one works with a lower-than-desired level of significance and loses power, or one adopts what is called a randomization rule. Few authors have recommended the latter course. Hence one should recognize that exact tests tend to give more conservative results than might otherwise be expected. For more discussion of this point, see Conover (1974).

A third limitation is that exact tests are usually focused on a single well-defined hypothesis. Thus they tend to be less useful when the study objectives include estimating several population parameters and detailed data analysis. It is possible to construct "exact intervals estimates," but this is somewhat complex and outside the scope of the present text. For more details on these interval estimates, the reader should refer to Cox (1970), Gart (1971), and Zelen (1971).

Section 8.1 begins with the hypergeometric distribution in the context of a test for independent observations. This distribution is generally similar to the binomial. The distribution is then used for the analysis of a 2×2 table (Section 8.2). The procedure is termed Fisher's Exact Test. Finally, the methodology is extended to several 2×2 tables employing what is known as the Mantel–Haenszel (1959) test (Section 8.3). The test was first presented in a slightly modified form by Cochran (1954) and its optimal properties demonstrated by Birch (1964). In epidemiology the test is sometimes termed the log rank test.

8.1 HYPERGEOMETRIC DISTRIBUTION

Case workers and census personnel are particularly concerned about the extent of coverage or case finding a particular screening technique attains. One method is based on the Chandrasaker–Deming (1949) methodology. The technique has also been used for estimating the size of wildlife populations. The estimation problem is discussed in a later section. Here we are more concerned with an underlying assumption.

The procedure usually begins by using some standard methodology to obtain a random sample of the population of interest. After noting a characteristic of the subjects or marking some of the subjects, they are returned to the population. At a later date another sample is obtained and the number of marked subjects is noted. It is usually assumed the two samples are collected independently. We examine this assumption in the course of developing the hypergeometric distribution.

Suppose that exactly 100 fish are in an experimental pond. In an initial sampling 15 fish are netted and marked. These are returned to the pond. After a few days a second netting yields 20 fish, of which 5 are marked. A natural question is whether the fish netted the first time are more or less likely to be netted than fish from the general population. Put differently, are the two samples really independent? To answer this question, we can build a model under the assumption of independence and determine the probability of obtaining 5 or more marked fish.

The data from the hypothetical experiment are shown in Table 8.1. There were 15 fish marked in the first sample, leaving 85 unmarked fish This is shown as the column totals. The sample was returned to the pond and another sample of 20 fish obtained. In the second sample of 20, 5 were marked and 15 were unmarked. This is shown in the first row of Table 8.1. Now if the samples were obtained independently, how likely is it that we would obtain 5 or more fish that were marked in a random sample of 20 fish? There are ${}_{100}C_{20}$ possible samples of size 20 that can be obtained from a population of size 100, and $({}_{15}C_5)({}_{85}C_{15})$ ways of obtaining 5 fish that are marked in both samples. Let Y, the number of fish that are marked in both samples, denote the random variable. It follows that $P(Y = 5) = ({}_{15}C_5)({}_{85}C_{15})/{}_{100}C_{20} = 0.101$. Similarly,

Table 8.1 Number of Fish Marked in Two Samples from a Pond Containing 100 Fish

Marked	Sample 1		
	Yes (1)	No (2)	Total (3)
Sample 2			
Yes	5	15	20
No	10	70	80
Total	15	85	100

$P(Y = 6) = 0.035$, $P(Y = 7) = 0.009$, $P(Y = 8) = 0.002$, $P(Y >= 9) < 0.001$. So $P(Y >= 5) = (0.101 + 0.035 + 0.009 + 0.002) = 0.147$. Thus there is a 14.7 percent chance of obtaining 5 or marked fish in sample of 20. Had 6 marked fish been obtained, one might question the assumption of independence.

The name of the distribution constructed in this example is the hypergeometric distribution. The general form is

$$P(Y = y) = \frac{({}_xC_y)({}_{n-x}C_{m-y})}{{}_nC_m} \qquad 0 < y < \min(x,m). \tag{8.1}$$

Here x is the number in the first sample or the number of subjects with a characteristic, m is the total number of subjects in the second sample, and y is the number in both samples with the characteristic. Finally, n is the total population size. For convenience we assume that $n \geq x + m$.

The hypergeometric distribution sums to 1. Without loss of generality, let $x \leq m$; then $\Sigma_{y=0}\, ({}_xC_y)({}_{n-x}C_{m-y})/({}_nC_x) = 1$, which follows from (2.21). Using the same procedure as for the binomial, we also obtain the mean and variance of Y:

$$E(Y) = \frac{mx}{n} \qquad \text{Var}(Y) = \frac{mx(n - x)(n - m)}{n^2(n - 1)}. \tag{8.2}$$

Notice that if we let $x/n = p$, then (8.2) is quite similar to the binomial:

$$E(Y) = mp \qquad \text{Var}(Y) = \frac{mp(1 - p)(n - m)}{n - 1}.$$

The final ratio in the variance is called the *finite population correction* (f.p.c.). Clearly, as n becomes large, the hypergeometric is close to the binomial. For example, we can compute the probability of obtaining exactly 5 marked fish in our sample of size 20 when $p = 0.85$ under a binomial model: $P(Y = 5) = {}_{20}C_5(0.15)^5(0.85)^{15} = 0.103$. We can also compare the mean and variance. The result is shown in Table 8.2. As might be expected, the large-sample approximation for the hypergeometric is similar to the binomial. For example, we can approximate $P(Y \geq 5)$ as follows:

$$z = \frac{(5 - 0.5) - 3}{\sqrt{2.06}} = \frac{1.50}{1.44} = 1.05.$$

Table 8.2 Mean and Variance of Binomial and Hypergeometric Distributions When $x = 15$, $m = 20$, and $n = 100$

Distribution		Hypergeometric	Binomial
$E(Y)$	Theoretical	mp	mp
	Observed	3	3
Var(Y)	Theoretical	$\dfrac{mp(1-p)(n-m)}{n^2(n-1)}$	$mp(1-p)$
	Observed	2.06	2.55

Here the 0.5 is subtracted as a continuity correction. Using $z = 1.05$ the probability of obtaining 5 or more marked fish is $P(Y \geq 5) = P(Z \geq 1.05) = 0.148$, which is somewhat larger than we obtained directly. This is partly due to the relatively small expectation, $mp < 5$.

8.2 FISHER'S EXACT TEST

Frequently, when studying a rare phenomenon the resulting data are quite unevenly distributed. One example is lung cancer among nonsmokers, because cigarette smoking is known to be the major cause of the disease. Correa et al. (1983) conducted a case-control study of lung cancer in order to assess the effects of passive smoking. One part of their results is shown in Table 8.3. Here nonsmokers are tabulated according to whether their spouse smoked. For the moment we will focus only on the males. There were only 8 lung cancer cases among ever-married, nonsmoking males. Moreover, only 28 spouses of the cases and controls combined smoked. Thus the minimum expected count under the hypothesis of homogeneity is $8(28)/188 = 1.19$, which is on the border of the requirement for employing the usual chi-square tests.

A test procedure can be developed from first principles. Assume that the case-control sample sizes are fixed by design. Since we have a nonrandom sample of the general population, the proportion of smoking spouses we happened to observe is of no interest. This proportion is termed a "nuisance parameter" and the usual procedure is to accept

Table 8.3 Spouse Smoking by Case-Control Status by Sex: Nonsmokers Who Have Been Married

		Spouse Smoked?		
Sex	Group	Yes	No	Total
Male	Case	2	6	8
	Control	26	154	180
	Total	28	160	188
Female	Case	14	8	22
	Control	61	72	133
	Total	75	80	155

Source: Correa et al. (1983).

the observed value as given. Thus the test is a conditional test of homogeneity.

These arguments lead to only one random variable, the number of smoking spouses among cases. How likely is it that exactly 2 are observed? Since there are 28 smoking spouses in our sample of 188, there are ${}_{188}C_{28} = 1.875 \times 10^{33}$ possible samples. Similarly, 2 of the 8 cases had smoking spouses. This can occur in $({}_8C_2)({}_{180}C_{26}) = 28(1.612 \times 10^{31})$ or 4.514×10^{32} ways. Hence the probability of the observed outcome given the marginal distributions is $P(Y = 2) = ({}_8C_2)({}_{180}C_{26})/({}_{188}C_{28}) = 0.241$. All other values are tabulated in Table 8.4, where y varies from 0 to 8.

Now under the null hypothesis (homogeneity) we are interested in values of y that are both large and small. The corresponding odds ratios are shown in the third column of Table 8.4. Thus small values of Y correspond to small odds ratios and large values to large odds ratios. The probability of observing $Y \leq 2$ is the sum of the first three probabilities or 0.901. Similarly, the probability of observing a significantly large odds ratio corresponds to observing that $Y > 2$ or 0.34. The associated p-value for the upper tail test is the latter value, so we must accept the hypothesis of homogeneity. The p-value for the correspond-

Table 8.4 Possible Response Distribution for Males in Table 8.3 Given the Observed Margins

Case (Y)	Control	Odds Ratio	$P(Y = y)$
0	28	0	0.268
1	27	0.81	0.392
2[a]	26	1.97	0.241
3	25	3.72	0.081
4	24	6.50	0.016
5	23	11.38	0.002
6	22	21.55	<0.001
7	21	52.33	<0.001
8	20	Undefined	<0.001

[a]Observed table.

ing two-tailed test is usually taken to be twice the minimum of the two p-values.

The general form of this test (Fisher's exact test) depends on Y following a hypergeometric distribution:

$$P(Y = y) = \frac{({}_{n_1}C_y)({}_{n_2}C_{x-y})}{{}_{(n_1+n_2)}C_x}, \tag{8.3}$$

where Y is the number of positive responses in the first sample of size n_1, n_2 is the second sample size, and x is the total number of positive responses.

The null hypothesis is that of homogeneity. For a one-tailed test it is rejected when $P[Y \leq y \text{ (observed)}] \leq a$: lower tail; or $P[Y \geq y$ (observed)$] \geq \alpha$: upper tail. For two-tailed alternatives, the hypothesis is rejected whenever

$$2 \min \{P[Y \leq y \text{ (observed)}], P[Y \geq y \text{ (observed)}]\} \leq \alpha.$$

The preceding is computationally straightforward, although tedious for moderately large sample sizes. One laborsaving device is simply to note which cell has the smallest observed value. Then using the cell as a starting point, compute $P\{Y \leq y[\min\text{ (observed)}]\}$. Using the relationship

$$P\{Y \geq y[\min\text{ (observed)}]\} = 1 + P\{Y = y\ [\min\text{ (observed)}]\} - P\{Y < y\ [\min\text{ (observed)}]\}$$

the other tail probability can be computed.

8.3 2 × 2 × *K* TABLES

The data for males in Table 8.3 do not provide significant evidence of a passive smoking relationship with lung cancer. A similar result holds for females. This is in spite of the respective odds ratios being 1.97 and 2.07. The similarity of the odd ratios suggests that the data can be combined to produce a single overall test of the association. Birch (1964) showed that in the absence of a three-variable interaction the uniformly most powerful test using the combined tables is the Mantel–Haenszel test.

The test statistic is computed by comparing the observed counts to what is expected under the hypothesis of homogeneity (or independence). It then uses a pooled estimate of variance under that hypothesis to generate a chi-square statistic. The *Mantel–Haenszel statistic* is given by

$$Q_c = \frac{[|\Sigma_k Y_k - \Sigma_k E(Y_k)| - 0.5]^2}{\Sigma_k V_c(Y_k)} \tag{8.4}$$

where

$$Y_k = \text{observed positive responses among cases at level } k$$

$$E(Y_k) = \frac{n_{1k}x_k}{n_{ik} + n_{2k}}$$

$$V_c(Y_k) = \frac{n_{1k}n_{2k}x_k(n_{1k} + n_{2k} - x_k)}{(n_{1k} + n_{2k})^2(n_{1k} + \mathrm{n}_{2k} - 1)}$$

$$n_{1k} = \text{number of cases in } k\text{th table}$$

$$n_{2k} = \text{number of controls in } k\text{th table}$$

$$x_k = \text{total positive responses in } k\text{th table.}$$

The subscript c emphasizes that this statistic is corrected for continuity. An uncorrected form was proposed by Cochran (1954) and is known as the *Birch–Cochran–Mantel–Haenszel statistic*:

$$Q = \frac{[\Sigma_k Y_k - \Sigma_k E(Y_k)]^2}{\Sigma_k (Y_k)} \qquad V(Y_k) = \frac{n_{1k} n_{2k} x_k (n_1 + n_{2k} - x_k)}{(n_{1k} + n_{2k})^3}. \tag{8.5}$$

To date no exhaustive comparison of the two has been conducted, although the former is more commonly reported.

The two procedures can be illustrated using the data in Table 8.3. Here $Y_1 = 2$, $Y_2 = 14$; $E(Y_1) = 1.191$, $E(Y_2) = 10.645$; $V_c(Y_1) = 0.976$ and $V_c(Y_2) = 4.745$. Hence $Q_c = [|16 - 11.836| - 0.5]^2/5.721 = 2.346$. Similarly, $V(Y_1) = 0.971$ and $V(Y_2) = 4.71$, so $Q = (16 - 11.836)^2/5.685 = 2.095$. Each when compared to a chi-square distribution, df $= 1$ yields a p-value > 0.10, so the association remains nonsignificant.

An alternative approach to these data is to fit a log-linear model. This would permit investigating whether the three-variable interaction is significant (it is not) and whether the data can be collapsed across sex. The latter requires that either sex and response or sex and group be conditionally independent. Each of the two approaches (log-linear modeling and Mantel–Haenszel) has advantages. The former assumes adequate sample sizes but permits examining the triple interaction and testing for collapsibility. It also requires considering the effect on the overall significance level of multiple tests. The latter approach can be used for smaller samples and preserves the overall significance level. The choice in any given study again depends on the original objectives.

8.4 COMPLEX RANDOM SAMPLES

The most important assumption for statistical analyses is probably that the observations were obtained through some type of random sampling. Usually, this means assuming that the observations are independently selected and they are homogeneously or identically distributed within subgroups. When the study is based on casually obtained samples, this assumption cannot be directly evaluated. However, good study design tends to reassure the analyst about whether the assumption is at least plausible. There are two important exceptions that ensure randomization: randomized experiments and complex probability samples. In the preceding chapters several examples of the former have been discussed. Complex probability samples are more difficult to analyze both conceptually and in practice.

Examples of complex probability samples include the U.S. National Health Survey (or at least its components), the Indian National Sample Survey, and the Current Population Survey, to name just a few. The key characteristic of these surveys is that each member of the sampled population has a known positive probability of being included in a given sample. Moreover, usually the sampled population corresponds closely to a well-defined, finite target population, and observations are selected without replacement. This means that the selections cannot be independent. In theory this dependency due to nonreplacement is unimportant, especially if both the sample and the target population are quite large. A more important source of dependency occurs when selection procedures attempt to exploit prior knowledge about the target population so as to increase precision and minimize cost. These practical considerations usually induce substantial dependencies among the observations that should be at least evaluated as part of the analysis.

The two most common ways in which dependencies are introduced is by stratification and clustering. Stratification means dividing the population into homogeneous subgroups. When this is done carefully it can have the effect of reducing variance in the estimates. However, it also means that the analysis must take account of often-differing sampling probabilities across the strata. In general, this is not a terribly difficult problem, although ignoring it completely can result in seriously biased estimates in some situations.

The second way that dependencies are introduced is by clustering observations and selecting the clusters at random rather than the individual observations. Usually, this causes some inflation of the variances, but is compensated for by reducing the overall cost of data collection. For large state or national surveys it is simply not feasible to use other than cluster samples at one or more stages of the survey.

The conceptual difficulty with analyzing these samples is that it is not always clear what traditional statistical inference, such as hypothesis testing and regression analysis, means in this context. For example, Cochran (1977, p. 39) points out that the hypothesis of equality of parameters for two populations is meaningless since the likelihood of exact equality is extremely small. Instead, we should think of the two populations as being samples from an infinite "superpopulation." The hypothesis is then about whether the populations were sampled from the *same* superpopulation. The implication is that for this type of analysis the fact that the populations are finite should be ignored.

The problem becomes more complex when stratification is considered. Generally, it is unclear whether the weights that reflect the sample design should be considered when using a regression model, especially when there appear to be no violations of the model assumptions after an analysis of the residuals. For an extended discussion of this point, see DuMouchel and Duncan (1983). Since stratification has been employed, it is clear that the assumptions *have* been violated. All residual analysis does is indicate how badly the assumptions have been violated. Moreover, residual analysis is always a subjective exercise; hence ignoring the weights completely is probably a questionable practice.

The finite population and stratification problems have at least been studied; however, the effects of clustering and the use of multiple sampling stages have been subject to only limited investigation. The only clear evidence comes from large-sample theory. Fuller (1975) shows that for a wide class of sample designs, the means for subgroups will follow a multivariate normal distribution as the samples become large. Based in part on this result, Koch, Freeman, and Freeman (1975) advocate using weighted least squares estimators for the analysis of complex sample surveys. These estimators and the associated test statistics are the same as those proposed by Grizzle et al. (1969) for the analysis of categorical data. The essential difference is the covariance matrix is estimated in such a way as to reflect the survey design rather than assuming a multinomial. It should be noted that the KFF approach also incorporates the Taylor Series approach to indirect variance estimation as a special case (Freeman et al. 1976).

Rather than recapitulate the underlying weighted least squares theory (Appendix 3A) we will illustrate the approach with three examples. The first percentages are compared across five groups. This is analogous to the traditional one-way analysis of variance. Next, mean birth weights are compared across six groups defined by two factors. Here the analogy is to two-way analysis of variance. Both of these examples use published estimates and standard errors only. In the third example it was possible to estimate the full set of variances and covariances, so the effect of the latter can be illustrated. In addition, the raw data were available, so a multinomial-based analysis is also conducted.

By implication, the other difficulty in survey analysis is the practical one of computational complexity. This has two parts: (1) availability and accessability of the data, and (2) complexity of estimation. Often, survey data are available only as estimates. When the detailed data are available, the data set may be so large that routine computations are terribly expensive. The latter problem is dramatically increased when the full covariance matrix is to be estimated. In the first two examples published estimates were used and the analysis was conducted using standard software (PROC REG) (SAS 1982b). The analysis could also be done using single-purpose software such as GENCAT (Landis et al. 1975). When the full data set was available, a special-purpose program for variance–covariance matrix estimation had to be written. The program GENCAT was then used to complete the analysis.

8.4.1 One-Way Analysis of Variance

The estimated percent of white adults in the United States (1962) needing to see a dentist by marital status is shown in Table 8.5 together with the sample-based standard errors (SE). These five groups are termed *domains* and correspond to the levels of the single factor "marital status." The lowest percent is for the never married and the highest percent is for the separated. Two questions are important:

1. Is there significant variation among the domains?
2. Which specific groups are different from the others?

There are two approaches to answering these questions. The first approach is more traditional and conservative. Here each group is compared to an overall estimate of the mean and using pairwise comparison tests, as was done in Section 2.9. A second approach is to estimate an analysis-of-variance model and then simplify it using a backwards elimination procedure on the estimated parameters. This approach depends on the parameterization selected. If this is done on substantive grounds, the analysis may have a somewhat arbitrary quality.

Table 8.5 Percent of White Adults Needing to See a Dentist by Marital Status[a]

Domain	Married	Widowed	Divorced	Separated	Never Married
Percent	37.8	42.0	42.6	62.7	31.8
SE	2.2	4.9	5.8	7.1	2.7
Married					
D_{ij}[b]	—	−4.2	−4.8	−24.9	6.0
$\mathrm{SE}(D_{ij})$[c]	—	5.37	6.20	7.43	3.48
Widowed	−0.78	—	−0.6	−20.7	10.2
			7.59	8.63	5.61
Divorced	−0.77	−0.08	—	−20.1	10.8
				9.17	6.40
Separated	−3.35	−2.40	−2.19	—	30.9
					7.60
Never married	1.72	−1.82	1.69	4.07	—

[a]Differences and standard errors above diagonal; z-test below the diagonal.
[b]$D_{ij} = P_i - P_j$.
[c]$\mathrm{SE}(D_{ij}) = (\mathrm{SE}^2(P_i) + \mathrm{SE}^2(P_j))^{1/2}$.
Source: National Center for Health Statistics (1973).

The first step is to estimate the overall mean from the data. This is the usual weighted least squares estimate:

$$\hat{p} = \frac{\Sigma_{i=1}^{5} \hat{p}_i/\hat{s}_i^2}{(\Sigma_{i=1}^{5} 1/\hat{s}_i)^2}, \tag{8.6}$$

where p_i is the estimated percentage and s_i its standard error for the ith domain, $i = 1, \ldots, 5$. In the example the estimate is

$$\hat{p} = \frac{37.8/4.84 + 42.0/24.01 + \cdots + 31.8/7.29}{1/4.84 + \cdots + 1/7.29}$$

$$= \frac{16.4315}{0.4350} = 37.77.$$

The test of no difference among domains (question 1) is best stated as the null hypothesis that the five domains are drawn from the same superpopulation. The correspond Neyman chi-square is

$$Q = \frac{\Sigma_{i=1}^{5} (\hat{p}_i - \hat{p})}{s_i}. \tag{8.7}$$

Since there are five domains and one estimate, the degrees of freedom are df = 5 − 1 = 4. The value for these data is

$$Q = \frac{(37.8 - 37.77)^2}{4.84} + \frac{(42.0 - 37.77)^2}{24.01} + \cdots = 18.6569.$$

This falls between the 99.5 and 99.95 percentiles of the chi-square (df = 4) distribution. Hence there is evidence of significant variation among the domains.

Answering the second question requires that we compare the domains two at a time. This implies ${}_5C_2 = 10$ comparisons. The overall significance level, a, for a two-tailed test is preserved by using the $[1 - a/2(10)] \times 100$ percentile of the appropriate distribution. For if $a = 0.05$, the $[1 - 0.05/20] \times 100 = 99.75$th percentile should be used.

The usual test statistic is the z-test:

$$z = \frac{p_i - p_j}{\sqrt{s_i^2 + s_j^2}}. \tag{8.8}$$

This follows a normal distribution for large samples, so the critical value for $\alpha = 0.05$ is 2.81. Similarly, if $\alpha = 0.10$, the percentile 99.5 and the critical value is 2.576. For example, comparing married adults to separated adults yields $z = (37.8 - 62.7)/(4.84 + 50.41)^{1/2} = -24.9/7.43 = -3.35$. This indicates that the two domains were sampled from different superpopulations; that is, the difference is significant.

The numerators and denominators of the z-tests are shown in the upper triangle of Table 8.5. The corresponding z-tests are shown in the lower triangle of Table 8.5. It is evident that the married and never-married percentages are significantly less than the separated percentage, but not different from each other. The widowed and divorced groups do show a significant difference from each other or the other three groups. The largest difference is −2.40, which is smaller in absolute value than either of the critical values. This somewhat fuzzy conclusion corresponds to the overlapping of confidence intervals, which is often seen in multiple comparison analysis.

A second approach begins by fitting a saturated model to the data and progressively eliminating nonsignificant parameters. As with any model-building procedure, this approach is somewhat ad hoc in that different analysts may identify different models which are equally appropriate. In the present case, one relevant model is based on comparing each of the ever-married groups to the never-married.

The reparameterized model can be written

$$E(p_1) = M + X_1B_1 + X_2B_2 + X_3B_3 + X_4B_4, \qquad (8.9)$$

where

$$\begin{aligned} X_1 &= 1 \text{ if } i = 1, \quad 0 \text{ otherwise} \\ X_2 &= 1 \text{ if } i = 2, \quad 0 \text{ otherwise} \\ X_3 &= 1 \text{ if } i = 3, \quad 0 \text{ otherwise} \\ X_4 &= 1 \text{ if } i = 4, \quad 0 \text{ otherwise.} \end{aligned}$$

The B_i are unknown parameters reflecting the difference between the ith group and the never-married group. Using weighted least squares the estimated model is

$$\begin{array}{ccccc} \hat{p}_1 = 31.8 & + \; 6.0X_1 & + \; 10.2X_2 & + \; 10.8X_3 & + \; 30.9X_4. \\ (2.7) & (3.5) & (5.6) & (6.4) & (7.6) \end{array}$$

Here the standard errors are shown in parentheses.

The simultaneous test that $B_1 = B_2 = B_3 = B_4 = 0$ yields the same value as before ($Q = 18.66$). In addition, testing each parameter individually, we find that the test statistic for B_3 yields $Q = 2.85$, which is the smallest of the four tests. Here the test statistic is the usual Wald test: $W = b_i/s_i$, where s_i is the standard error of b_i. The next step is to fit a new model deleting B_3.

At the second step, B_1 yields the smallest Q test. Continuing to delete and test, we find that only B_4 is retained. The sequence is

Table 8.6 Sequential Tests for Model Building Using the Data in Table 8.5

Test (Deleted Terms)	df	Q	p-Value	Δdf	ΔQ	p-Value
$B_3 = 0$	1	2.85	0.09	1	2.85	0.09
$B_1 = B_3 = 0$	2	4.38	0.11	1	1.53	0.22
$B_1 = B_2 = B_3 = 0$	3	5.74	0.12	1	1.36	0.24
$B_1 = B_2 = B_3 = B_4 = 0$	4	18.66	<0.01	1	12.91	<0.01

displayed in Table 8.6. Here the Q values are the residual tests for the corresponding models. The fourth column, ΔQ, is the change in residual due to the most recently deleted term. It is identical to the Wald test in the model with the term included. Notice that these sequential tests are in fact additive.

The backwards elimination of parameters yields $p_1 = 36.6 + 26.1X_4$ as a final model. Thus only the separated group is found to be significantly different from the others. This can be reported in terms of fitted values. For the four groups 36.6 ± 1.6 percent need to see a dentist, but among separated persons the percentage is 62.7 ± 7.1.

An alternative model selection procedure would be to test all possible combinations of the parameters being equal to zero. There are four tests of $B_i = 0$, six tests of $B_i = B_j = 0$, four tests of $B_i = B_j = B_k = 0$, and one test of $B_1 = B_2 = B_3 = B_4 = 0$. In general, for I parameters there are $2^I - 1$ possible tests. Obviously, as I becomes large, this can be somewhat cumbersome.

8.4.2 Two-Way Analysis of Variance

The problem of testing the effect of a factor on a response while controlling for a second factor leads immediately to two-way analysis of variance. For example, suppose that one wishes to investigate the association between a mother's working and her child's birth weight. It is suspected that birth weight is related to family income, so this must be controlled in the analysis. As an example, data on mean birth weight by employment status and family income are shown in Table 8.7. Descriptively, birth weight increases with family income and for unemployed mothers. To determine whether this is "real," we must conduct statistical tests.

Table 8.7 Mean Birth Weight by Family Income and Mother's Employment Status, United States, 1963

Family Income ($1000s) ($i$)	Mother's Employment (j)	Birth Weight (g)		Design Variable		
		Mean	S.E.	X_1	X_2	X_3
3–<5	Yes (1)	3230	24.36	1	0	1
(1)	No (2)	3290	17.96	1	0	0
5–<7	Yes (1)	3280	22.11	0	1	1
(2)	No (2)	3320	18.08	0	1	0
7+	Yes (1)	3280	21.15	0	0	1
(3)	No (2)	3360	18.36	0	0	0

Source: NCHS (1973).

Since there are two categorical factors it seems natural to fit an analysis-of-variance model to the birth weights. The estimates are derived from a sample survey, but only the standard errors were estimated. The squared standard errors can be used for the weights in a weighted least squares program, as was done in the example in Section 8.4.1. The two-factor ANOVA model can be written

$$E(\overline{Y}_{ij}) = M + I_1X_1 + I_2X_2 + EX_3, \tag{8.10}$$

where

$X_1 = 1$ if $i = 1$, 0 otherwise

$X_2 = 1$ if $i = 2$, 0 otherwise

$X_3 = 1$ if $j = 1$, 0 otherwise.

Here $\overline{Y}_{ij}$ is the mean birth weight for the ith income group and jth employment status. The design variables are used to contrast low to high income (X_1, X_2) and employed to unemployed (X_3), as shown in Table 8.7.

The resulting analysis-of-variance tests are shown in Table 8.8. The error line shows the hypothesis of no interaction between income and employment can be accepted ($Q = 1.00$, df $= 2$). That is, the

Table 8.8 Analysis of Variance for Mean Birth Weight by Family Income and Mother's Employment

Hypothesis	df	Q	p-Value
Model	3	21.83	< 0.01
Income	2	9.65	0.01
Employment	1	13.24	< 0.01
Error (interaction)	2	1.00	0.61
Total	5	22.83	< 0.01

Estimated Model

$$E(Y_{ij}) = \underset{(15.6)}{3352} - \underset{(19.9)}{61.4X_1} - \underset{(19.7)}{23.4X_2} - \underset{(16.6)}{60.3X_3}$$

$X_1 = 1$ if family income $3 - <5$ $(i = 1)$, 0 otherwise
$X_2 = 1$ if family income $5 - <7$ $(i = 2)$, 0 otherwise
$X_3 = 1$ if mother employed, 0 otherwise

data fit the hypothesized model. Specific hypotheses can be tested by using contrast matrices, as discussed in Chapter 4. For example, the *model hypothesis* is given by

$$H_0\colon \ \mathbf{CB} = 0, \qquad \mathbf{C} = \begin{bmatrix} 0 & 1 & 0 & 0 \\ 0 & 0 & 1 & 0 \\ 0 & 0 & 0 & 1 \end{bmatrix} \qquad \mathbf{B} = \begin{bmatrix} M \\ I_1 \\ I_2 \\ E \end{bmatrix}$$

This yields a significant Neyman chi-square (Wald test) ($Q = 21.83$, df $= 1$). Similarly, the income ($Q = 9.65$, df $= 2$) and employment ($Q = 13.24$, df $= 1$) factors have a statistically significant effect on mean birth weight.

The magnitude and direction of the effects can be determined from the estimated model (bottom of Table 8.8). Since X_1 corresponds to the lowest-income group, its coefficient indicates a reduction in birth weight by 61.4 g. The middle level of income reduces birth weight by only 23.4 g, which is not significant ($23.4/19.7 = 1.2$). Finally, employment reduces birth weight by 60.3 g. While the language implies causality, it should be emphasized that these are strictly statistical associations. In summary, being in the lowest-income group or employed is associated with reduced birth weight, and these effects are additive.

8.4.3 Comparison of Complex to Simple Sampling

The examples in Sections 8.4.1 and 8.4.2 are based on published data for which only estimates and standard errors were available. The estimated prevalence of uncontrolled hypertension is shown in Table 8.9. The estimates and sampling procedures are discussed in detail by Freeman et al. (1983). In addition, the complete covariance matrix is shown. The interdomain covariances are above the diagonal and the corresponding correlations are below the diagonal. As can be seen, the estimates for one domain are correlated with those of another. For example, the estimates for males and females under 50 are negatively correlated ($r = -0.16$), but over 50 they are positively correlated ($r = 0.30$). The precise source of the correlation is unknown, but it is thought to be due to the sample design employing clusters of housing units. The estimated variances are shown on the diagonal. A weighted least squares analysis of these data would utilize the complete covariance matrix as input to a program such as GENCAT (Landis et al. 1976).

For comparison the "raw" or unweighted data are shown in the last three columns of Table 8.9. The variances are based on the usual pq/n estimator for a binomial. The estimated prevalences are quite

Table 8.9 Percent Uncontrolled Hypertension by Age and Sex for Complex and Simple Random Samples

Group		Complex Sample Estimates					Simple Sample Estimates		
Age	Sex	Percent	Covariance Matrix				N	Percent	Variance
<50	Male	90.8	6.4	−2.4	−1.7	0.3	113	90.3	7.8
	Female	79.3	−0.16	35.6	2.1	−8.9	49	77.6	35.5
50+	Male	80.8	−0.25	0.13	7.3	3.6	192	80.2	8.3
	Female	66.1	0.02	−0.33	0.30	20.8	251	64.9	9.1

similar indicating that the design was approximately self-weighting. Interestingly, for three of the four domains the variances are also similar. However, for females at least age 50, the complex sampling variance estimate is more than twice what is obtained from the binomial estimator. The former is what should be reported. Because of the similarity between the estimates, one would expect the resulting analyses to be similar.

The analysis is based on a two-factor ANOVA model, as in Section 8.4.2:

$$E(P_{ij}) = M + AX_1 + SX_2, \tag{8.11}$$

where $X_1 = 1$ if age < 50 $(i = 1)$, 0 otherwise, $X_2 = 1$ if male $(j = 1)$, 0 otherwise. Here A compares young to old and S compares males to females. The resulting test statistics are shown in Table 8.10 and the estimated model parameters in Table 8.11. As was expected, the complex and simple random sampling approaches yield similar results. This suggests that for this particular survey, the simple random sampling assumptions may not be inappropriate, at least approximately. However, for other surveys, particularly non-self-weighting designs, the weighted least squares method employing the complete covariance matrix should be examined.

Table 8.10 Analysis of Variance for Complex and Simple Random Sampling Models

		Complex Sample		Simple Random	
Hypothesis	df	Q	p-Value	Q	p-Value
Model	2	27.58	<0.01	38.24	<0.01
Age	1	8.37	<0.01	9.76	<0.01
Sex	1	21.36	<0.01	17.07	<0.01
Error	1	0.11	0.74	0.11	0.74
Total	3	27.70	<0.01	38.35	<0.01

Table 8.11 Estimated Model Parameters and Standard Errors for Complex and Simple Random Sampling

	Complex		Simple	
Parameter	Estimate	SE	Estimate	SE
M	67.0	3.7	65.3	2.8
Y	10.7	3.7	10.7	3.4
S	13.6	2.9	14.5	3.5

References

American Heritage Dictionary of the English Language. W. Morris ed. (1969). Boston: Houghton Mifflin.

Arbous, A. G., and Kerrich, J. E. (1951). Accident statistics and the concept of accident proneness, *Biometrics 7,* 340–432.

Baker, R. J., and Nelder, J. A. (1978). *The GLIM System Release 3: Generalized Linear Interactive Modelling Manual*. Oxford: Numerical Algorithms Group.

Bartlett, M. S. (1935). Contingency table interactions, *J. R. Stat. Soc. Sup.,* 248–252.

Berkson, J. (1944). Application of the logistic function to bio-assay, *J. Am. Stat. Assoc. 39,* 357–365.

———. (1953). A statistically precise and relatively simple method of estimating the bio-assay with quantal response, based on the logistic function, *J. Am. Stat. Assoc. 48,* 565–599.

———. (1955). Maximum likelihood and minimum X^2 estimates of the logistic function, *J. Am. Stat. Assoc. 50,* 130–162.

Bhapkar, V. P. (1966). A note on the equivalence of two criteria for hypotheses in categorical data, *J. Am. Stat. Assoc. 61,* 228–235.

Birch, M. W. (1964). The detection of partial association, I: the 2 × 2 case, *J. R. Stat. Soc. Ser. B. 26,* 313–324.

Bishop, Y. M. M. (1969). Full contingency tables, logits, and split contingency tables, *Biometrics 25,* 383–399.

Bishop, Y. M. M., Fienberg, S. E., and Holland, P. W. (1975). *Discrete Multivariate Analysis: Theory and Practice,* Cambridge, Mass.: The MIT Press.

Bross, I. D. J. (1958). How to use ridit analysis, *Biometrics 14,* 18–38.

Brown, M. B. (1976). Screening effects in multidimensional contingency tables, *App. Stat. 25,* 37–46.

Caussinus, H. (1966). Contribution a l'analyse statistique des tableaux des correlation, *Ann. Fac. Sci U. Toulouse 29,* 77–182.

Chandrasekar, C., and Deming, W. Edward (1949). On a method of estimating birth and death rates and the extent of registration, *J. Am. Stat. Assoc. 44,* 101–115.

Chagnon, N. A. (1977). *Yanomamö, The Fierce People,* New York: Holt, Rinehart, and Winston.

Chiang, Chin Long (1968). *Introduction to Stochastic Processes.* New York: John Wiley and Sons, Inc.

Cochran, W. G. (1954). Some methods for strengthening the common X^2 tests, *Biometrics 10,* 417–451.

———. (1977). *Sampling Techniques,* 3rd ed., New York: John Wiley & Sons, Inc.

Colton, T. (1974). *Statistics in Medicine.* Boston: Little, Brown.

Conover, W. J. (1974). Some reasons for not using the Yates continuity correction on 2 × 2 contingency tables, *J. Am. Stat. Assoc. 69,* 374–378.

Cornfield, J. (1951). A method of estimating comparative rates from clinical data: Applications to cancer of the lung, breast and cervix, *J. Nat. Cancer Inst. 11,* 1269–1275.

Correa, D., Pickle, L. W., Fontham, E., Lin, Y., and Haenszel, W. (1983). Passive smoking and lung cancer, *Lancet 8350,* 595–597.

Cox, D. R. (1970). *The Analysis of Binary Data.* London: Methuen & Company Ltd.

Cox, D. R., and Oakes, D. (1984). *Analysis of Survival Data.* London: Chapman & Hall Ltd.

Cox, D. R., and Snell, E. J. (1968). A general definition of residuals (with Discussion), *J. R. Stat. Soc. Ser. B 30,* 248–275.

Cramer, H. (1946). *Mathematical Methods of Statistics.* Princeton: Princeton U. Press.

Darroch, J. N. (1976). No-interaction in contingency tables, *Proc. 9th International Biometric Conference,* 296–316.

D'Atri, D. A., Freeman, D. H., Ostfeld, A. M., Papke, E., Piorun, K., and Sardinas, A. (1979). Plan and operation of the Connecticut Hypertension Survey, *Am. Stat. Assoc.: Surv. Res. Meth. Sect.*, 179–181.

Dixon, W. J. (1981). *BMDP Statistical Software 1981*. Berkeley: Univ. of Calif. Press.

Draper, N., and Smith, H. (1981). *Applied Regression Analysis*, 2nd ed. New York: John Wiley and Sons, Inc.

DuMouchel, W. H., and Duncan, G. J. (1983). Using sample survey weights in multiple regression analyses of stratified samples, *J. Am. Stat. Assoc. 78*, 535–543.

Dyke, G. V., and Patterson, H. D. (1952). Analysis of factorial arrangements when the data are proportions, *Biometrics 8*, 1–12.

Everitt, B. S. (1977). *The Analysis of Contingency Tables*. London: Chapman & Hall Ltd.

Fienberg, S. E. (1980). *The Analysis of Cross-Classified Categorical Data*, 2nd ed. Cambridge, Mass.: MIT Press.

———, and Larntz, K. (1976). Loglinear representation for paired and multiple comparisons models, *Biometrika 35*, 479–481.

Finney, D. J. (1972). *Probit Analysis*, 3rd ed. London: Cambridge University Press.

Fischer, R. A. (1922). On the interpretation of X^2 from contingency tables and the calculation of P, *J. R. Stat. Soc. 85*, 87–94.

Fisz, M. (1963). *Probability Theory and Mathematical Statistics*. New York: John Wiley & Sons, Inc.

Fleiss, J. L. (1981). *Statistical Methods for Rates and Proportions*, 2nd ed. New York: John Wiley & Sons, Inc.

Forthofer, Ron N., Lehnen, Robert G. (1981). *Public Program Analysis: A New Categorical Data Approach*. Belmont, Calif.: Lifetime Learning Publications.

Freeman, D. H. (1981). The analysis of twice classified data, *American Statistical Association: 1981 Proceedings of the Social Statistics Section*, 178–182.

Freeman, D. H. (1983). Graphical representation of multiway contingency tables: alternative measures of association. *Am. Stat. Assoc.: 1983 Proceedings of the Social Statistics Section*, 544–548.

Freeman, D. H., D'Atri, D. A., Hellenbrand, K., Ostfeld, A. M., Papke, E., Piorun, K., Richards, V. A., and Sardinas, A. (1983). The prevalence distribution of hypertension: Connecticut Adults 1978–1979, *J. Chron. Dis. 36:2*, 171–181.

Freeman, D. H., Freeman, J. L., Brock, D. B., and Koch, G. G. (1976). Strategies in the multivariate analysis of data from complex surveys II: an application to the United States National Health Interview Survey, *Int. Stat. Rev. 44:3,* 317–330.

Freeman, D. H., and Jekel, J. F. (1980). Table selection and log-linear models, *J. Chron. Dis.,* 513–524.

Fuller, W. A. (1975). Regression analysis for sample survey, *Sankhya Ser. C 37,* 117–132.

Gart, J. J. (1971). The comparison of proportions: a review of significance tests, confidence intervals, and adjustments for stratification, *Rev. Int. Stat. Inst. 39,* 148–169.

Goodman, Leo A. (1978). *Analyzing Qualitative/Categorical Data: Log-Linear Models and Latent-Structure Analysis,* ed. by Jay Magidson, Cambridge, Mass.: Abt Books.

Goodman, L. A., and Kruskal, W. A. (1979). *Measures of Association for Cross-Classifications.* New York: Springer-Verlag.

Graham, S., and Shotz, W. (1979). Epidemiology of cancer of the cervix in Buffalo, New York, *J. of the Nat. Cancer Inst. 63:1,* 23–27.

Grizzle, James E., Starmer, C. Frank, and Koch, Gary G. (1969). Analysis of categorical data by linear models, *Biometrics 25,* 489–504.

Gross, Alan J., and Clark, Virginia A. (1975). *Survival Distributions: Reliability Applications in the Biomedical Sciences.* New York: John Wiley & Sons, Inc.

Guttman, L. (1941). An outline of the statistical theory of prediction, Supp. B-1, in Horst, Paul, et al. (eds.). *The Prediction of Personal Adjustment, 48.* New York: Social Science Research Council, 253–318.

Haberman, S. J. (1978, 1979). *Analysis of Qualitative Data.* New York: Academic Press, Inc.

Halperin, M., Blackwelder, W. C., and Verter, J. I. (1971). Estimation of the multivariate logistic risk function: A comparison of the discriminant function and maximum likelihood approaches, *J. Chron. Dis. 24,* 125–158.

Harrell, F. (1980). The logist procedure, *SAS Supplemental Library User's Guide,* SAS Institute, Cary, NC, 83–102.

Harris, B. (1966). *Theory of Probability.* Reading, Mass.: Addison-Wesley Publishing Company, Inc.

Higgens, J. E., and Koch, G. G. (1977). Variable selection and generalized chi-square analysis of categorical data applied to a large cross-sectional occupational health survey, *Int. Stat. Rev. 45,* 51–62.

Holford, T. R., White, C., and Kelsey, J. L. (1978). Multivariate analysis for matched case-control studies, *Am. J. Epidem. 108(3),* 245–256.

Hypertension Detection and Follow-up Program Cooperative Group (1982). The Effect of treatment on mortality in "mild" hypertension, *N. Eng. J. Med. 307,* 976–980.

Imrey, P. B., Koch, G. G., and Johnson, W. D. (1976). An incomplete contingency table approach to paired-comparison experiments, *J. Am. Stat. Assoc. 71,* 614–623.

Imrey, P. B., Koch, G. G., and Stokes, M. E. (1981). Categorical data analysis: Some reflections on the log linear model and logistic regression. Part I: Historical and methodological overview, *Int. Stat. Rev. 49,* 265–283.

———. (1981). Categorical data analysis: Some reflections on the log linear model and logistic regression. Part II: data analysis, *Int. Stat. Rev. 50,* 35–63.

Interstudy (1980). July 1980 survey results: HMO enrollment and utilization in the U.S. Excelsior, Minnesota: Interstudy.

Iverson, G. R. (1979). Decomposing chi-square: a forgotten technique, *Sociol. Meth. and Res. 8,* 143–157.

Jekel, J. F., Freeman, D. H., and Meigs, J. W. (1978). A study of trends in upper arm soft tissue sarcomas in the State of Connecticut following the introduction of alum-adsorbed allergenic extract, *Annals of Allergy 40:1,* 28–31.

Johnson, Norman L., and Kotz, Samuel (1969). *Discrete Distributions.* Somerset, N.J.: John Wiley & Sons, Inc.

Kalbfleisch, J. D., and Prentice, R. L. (1980). *The Statistical Analysis of Failure Time Data.* New York: John Wiley & Sons, Inc.

Kendall, M. G. (1949). Rank and product-moment correlation, *Biometrika 36,* 177–193.

Kleinbaum, David G., and Kupper, Lawrence L. (1978). *Applied Regression Analysis and Other Multivariate Methods.* Belmont, Calif.: Wadsworth Publishing Company, Inc.

Kleinbaum, David G., Kupper, Lawrence L., and Morgenstern, Hal (1982). *Epidemiologic Research: Principles and Quantitative Methods.* Belmont, Calif.: Wadsworth Publishing Company, Inc.

Koch, G. G. (1975, 1977). Lecture notes.

Koch, G. G., Freeman, D. H., and Freeman, J. L. (1975). Strategies in the multivariate analysis of data from complex surveys, *Int. Stat. Rev. 43:1,* 59–78.

Koch, G. G., Landis, J. R., Freeman, J. L., Freeman, D. H., and Lehnen, R. G. (1977). A general methodology for the analysis of experiments with repeated measurement of categorical data, *Biometrics 33,* 133–158.

Lancaster, H. C. (1951). Complex contingency tables treated by the partition of X^2, *J. R. Stat. Soc. Ser. B 13,* 242–249.

Landis, J. R., Stanish, W. M., Freeman, J. L., Koch, G. G. (1976). A computer program for the generalized chi-square analysis of categorical data using weighted least squares (GENCAT). *Comp. Prog. Biomed. 6,* 196–231.

Lee, E. T. (1980). *Statistical Methods for Survival Data Analysis.* Belmont, Calif.: Lifetime Learning Pub.

Lehman, E. L. (1975). *Nonparametrics: Statistical Methods Based on Ranks.* San Francisco: Holden-Day Co.

Levy, P. S., and Lemeshow, S. (1980). *Sampling for Health Professionals.* Belmont, Calif.: Lifetime Learning Pub.

Lombard, H. L., and Doering, C. R. (1947). Treatment of the fourfold table by partial association and partial correlation as it relates to public health problems, *Biometrics 3,* 123–128.

Magidson, Jay (1978). An illustrative comparison of Goodman's approach to logic analysis with dummy variable regression analysis, in L. A. Goodman, *Analyzing Qualitative/Categorical Data,* Cambridge, Mass.: Abt Books, 27–54.

Mantel, N., and Haenszel, W. (1959). Statistical aspects of the analysis of data from retrospective studies of disease, *J. Nat. Cancer Inst. 22,* 719–748.

Maxwell, A. E. (1961). *Analyzing Quantitative Data.* London: Methuen.

National Center for Health Statistics (1973). *Manual on Standards for Reviewing Statistical Reports.* Unpublished preliminary draft, cited by Koch *et al.* 1975.

———. (1976). Annual summary for the United States 1975, *Monthly Vital Statistics Report, 24:13.* HRA Pub. No. 76-1120, Public Health Service, Rockville, MD, Tables 1, 2.

———. (1980). Annual summary for the United States 1979, *Monthly Vital Statistics Report, 28:13.* DHHS Pub. No. 81-1120, Public Health Service, Hyattsville, MD, Tables 1, 2.

———. (1983). Births, marriages, divorces and deaths: United States 1982. *Monthly Vital Statistics Report, 31:12.* DHHS Pub. No. 83-1120, Public Health Service, Hyattsville, MD, Table 1.

——— (Graves, E., and Haupt, B. J.) (1983). Utilization of short stay hospitals, United States 1981, Annual Summary. *Vital and Health Statistics,* 13:72, DHHS Pub. No. (PHS) 83-1733, Public Health Service, Washington, D.C.

Nelder, J. A., and Wedderburn, R. W. M. (1972). Generalized linear models, *J. R. Stat. Soc. A 135,* 370–84.

Nerlove, M., and Press, S. J. (1973). *Univariate and Multivariate Log-linear and Logistic Models.* Santa Monica, Calif.: Rand Corp. Technical Report R-1306-EDA/NIH.

Neyman, J. (1949). Contribution to the theory of the X^2 test, *Proc. First Berkeley Symp. on Mathematical Statistics and Probability.* ed. by J. Neyman, Berkeley, Calif.: Univ. of Calif. Press, 230–273.

Nie, N. H., Hull, C. H., Jenkins, J. G., Steinbrenner, K., Bent, D. H. (1975). *SPSS: Statistical Package for the Social Sciences,* 2nd ed. New York: McGraw-Hill Book Company.

Norton, H. W. (1945). Calculation of chi-square for complex contingency tables, *J. Amer. Stat. Assoc. 40,* 251.

Novitski and Kimball (1958). *Amer. J. Hum. Gen. 10,* 268.

Ostfeld, Adrian M. (1980). Lecture notes.

Pearson, K. (1900). On the criterion that a given system of deviation from the probable in the case of a correlated system of variable is such that it can be reasonable, supposed to have arisen from random sampling, *Phil. Mag. 1, Ser. 5, Vol. 50,* 157–175.

Pearson, K. (1904). On the theory of contingency and its relation to association and normal correlation, *Draper's Co. Res. Mem. Biometric Ser. 1* Reprinted (1948) in Karl Pearson's Early Papers, Cambridge: Cambridge University Press.

Placket, R. L. (1974). *The Analysis of Categorical Data.* High Wycombe, Buckinghamshire England: Charles Griffin & Company Ltd.

Poisson, Simeon Denis (1837). Recherches sur la probabilité des jugements en matière criminelle et matière civil, in *Précédées des règles générales du calcul des probabilités.* Paris: Bachelier, Imprimerie-Librairie pour les Mathématiques la Physique.

Rao, C. R. (1973). *Linear Statistical Models,* 2nd. ed. New York: John Wiley & Sons, Inc.

Roy, S. N., and Kastenbaum, M. A. (1956). On the hypothesis of no "interaction" in a multi-way contingency table. *Ann. Math. Stat. 27,* 749–757.

SAS Institute Inc. (1982a). *SAS User's Guide: Basics, 1982 Edition,* Cary, N.C.: SAS Institute.

SAS Institute Inc. (1982b). *SAS User's Guide: Statistics, 1982 Edition,* Cary, N.C.: SAS Institute.

Scudder, R., and Anderson, C. A. (1954). Migration and vertical occupational mobility, *Am. Soc. Rev. 19,* 329–334.

Selvin, S. (1977). A further note on the interpretation of ridit analysis, *Am. J. Epidem. 105,* 16–20.

Senie, R. T., Rosen, P. P., Lesser, M. L., and Kinne, D. W. (1981). Breast self-examinations and medical examination related to breast cancer stage, *Am. J. Pub. Health,* 583–590.

Shapiro, S. H. (1982). Collapsing contingency tables—a geometric approach, *Am. Stat. 36,* 43–46.

Sheps, M. C. (1959). An examination of some methods of comparing several rates or proportions, *Biometrics 15,* 87–97.

Shyrock, H., and Siegel, J. (1973). *Methods and Materials of Demography.* Washingon, D.C.: U.S. Government Printing Office.

Simpson, E. H. (1951). The interpretation of interaction in contingency tables, *J. R. Stat. Soc. 13:2,* 238–241.

Snedecor, G. W., and Cochran, W. G. (1967). *Statistical Methods,* 6th ed. Ames, Iowa: Iowa State Univ. Press.

Snedecor, G. W., and Irwin, M. R. (1933). *Iowa State J. Sci. 8:75,* as Ames, Iowa: Iowa State Univ. Press.

Snedecor, G. W., and Cochran, W. G. (1984). *Statistical Methods,* 7th ed. Ames, Iowa: Iowa State Univ. Press.

Snedecor, G. W., and Irwin, M. R. (1933). *Iowa State J. Sci. 8:75,* as reported by Snedecor and Cochran (1967).

Somers, D. H. (1962). A new asymmetric measure of association for ordinal variables, *Am. Sociol. Rev. 27,* 799–811.

Stern, E., Misczyncki, M., Greenland, S., Damus, K., Coulson, A. (1977). Pap testing and hysterectomy prevalence: a survey of communities with high and low cervical cancer rates, *Am. J. of Epidem. 106,* 296–305.

Truett, J., Cornfield, J., and Kannel, W. (1967). A multivariate analysis of the risk of coronary heart disease in Framingham, *J. Chron. Dis. 20,* 511–524.

U.S. Bureau of the Census (1943a). *Sixteenth Census of the United States.*

U.S. Bureau of the Census (1943b). *Vital Statistics of the United States 1900–1940.*

Upton, G. J. G. (1978). *The Analysis of Cross-Tabulated Data.* New York: John Wiley & Sons, Inc.

Wald, A. (1943). Tests of statistical hypotheses concerning several parameters when the number of observations is large, *Transact. Am. Math. Soc. 54,* 426–482.

Walker, S. H., and Duncan, D. B. (1967). Estimation of the probability of an event as a function of several independent variables, *Biometrika 54,* 167–179.

Walter, S. D., and Holford, T. R. (1978). Additive, multiplicative, and other models for disease risk, *Am. J. of Epidem. 108,* 341–346.

Weissman, M., and Myers, J. (1978). Affective disorders and social class in an U.S. urban community, *Arch. Gen. Psych. 35,* 1304–1311.

Wermuth, N. (1976). Exploratory analyses of multidimensional contingency tables, in *Proc. 9th International Biometrics Conference, V. I,* 279–295.

West, E. D. (1969). The signed rank (Wilcoxon) test, *Lancet 1,* 526. cited by Colton (1974).

Whittemore, A. S. (1978). Collapsibility of multidimensional contingency tables, *J. R. Stat. Soc. Ser. B 40,* 328–340.

Wexler, L. M. (1979). Long-range prediction of change and stability in adult mental health, Ph.D. Dissertation, Yale University.

Winsor, C. P. (1948). Factorial analysis of a multiple dichotomy, *Hum. Bio. 20,* 195–204.

Yates, F. (1934). Contingency tables involving small numbers and the X^2 test, *J. R. Stat. Soc. Sup. I: 2,* 217–235.

Yates, F. (1948). The analysis of contingency tables with groupings based on quantitative characteristics, *Biometrika 35,* 176–181.

Yates, F. (1960). *Sampling Methods for Censuses and Surveys,* 3rd. ed. London: Griffin.

Yule, G. U. (1900). On the association of attributes in statistics, *Phil. Trans. Ser. A 194,* 257–319.

Zelen, M. (1971). The analysis of several 2 × 2 contingency tables, *Biometrika 58,* 128–138.

Appendix

Table A.1 Critical Values of the Chi-Square Distribution

	cumulative probability, γ					
f	0.005	0.010	0.025	0.050	0.100	0.250
1	-	-	0.001	0.004	0.016	0.102
2	0.010	0.020	0.051	0.103	0.211	0.575
3	0.072	0.115	0.216	0.352	0.584	1.213
4	0.207	0.297	0.484	0.711	1.064	1.923
5	0.412	0.554	0.831	1.145	1.610	2.675
6	0.676	0.872	1.237	1.635	2.204	3.455
7	0.989	1.239	1.690	2.167	2.833	4.255
8	1.344	1.646	2.180	2.733	3.490	5.071
9	1.735	2.088	2.700	3.325	4.168	5.899
10	2.156	2.558	3.247	3.940	4.865	6.737
11	2.603	3.053	3.816	4.575	5.578	7.584
12	3.074	3.571	4.404	5.226	6.304	8.438
13	3.565	4.107	5.009	5.892	7.042	9.299
14	4.075	4.660	5.629	6.571	7.790	10.165
15	4.601	5.229	6.262	7.261	8.547	11.037
16	5.142	5.812	6.908	7.962	9.312	11.912
17	5.697	6.408	7.564	8.672	10.085	12.792
18	6.265	7.015	8.231	9.390	10.865	13.675
19	6.844	7.633	8.907	10.117	11.651	14.562
20	7.434	8.260	9.591	10.851	12.443	15.452
21	8.034	8.897	10.283	11.591	13.240	16.344
22	8.643	9.542	10.982	12.338	14.041	17.240
23	9.260	10.196	11.689	13.091	14.848	18.137
24	9.886	10.856	12.401	13.848	15.659	19.037
25	10.520	11.524	13.120	14.611	16.473	19.939
26	11.160	12.198	13.844	15.379	17.292	20.843
27	11.808	12.879	14.573	16.151	18.114	21.749
28	12.461	13.565	15.308	16.928	18.939	22.657
29	13.121	14.256	16.047	17.708	19.768	23.567
30	13.737	14.953	16.791	18.493	20.599	24.478
35	17.192	18.509	20.569	22.465	24.797	29.054
40	20.707	22.164	24.433	26.509	29.051	33.660
45	24.311	25.901	28.366	30.612	33.350	38.291
50	27.991	29.707	32.357	34.764	37.689	42.942
55	31.735	33.570	36.398	38.958	42.060	47.610
60	35.534	37.485	40.482	43.188	46.459	52.294
65	39.383	41.444	44.603	47.450	50.883	56.990
70	43.275	45.442	48.758	51.739	55.329	61.698
75	47.206	49.475	52.942	56.054	59.795	66.417
80	51.172	53.540	57.153	60.391	64.278	71.145
85	55.170	57.634	61.389	64.749	68.777	75.881
90	59.196	61.754	65.647	69.126	73.291	80.625
95	63.250	65.898	69.925	73.520	77.818	85.376
100	67.328	70.065	74.222	77.929	82.358	90.133
120	83.852	86.923	91.573	95.705	100.624	109.220

	cumulative probability, γ					
f	0.750	0.900	0.950	0.975	0.990	0.995
1	1.323	2.706	3.841	5.024	6.635	7.879
2	2.773	4.605	5.991	7.378	9.210	10.597
3	4.108	6.251	7.815	9.348	11.345	12.838
4	5.385	7.779	9.488	11.143	13.277	14.860
5	6.626	9.236	11.070	12.833	15.086	16.750
6	7.841	10.645	12.592	14.449	16.812	18.548
7	9.037	12.017	14.067	16.013	18.475	20.278
8	10.219	13.362	15.507	17.535	20.090	21.955
9	11.389	14.684	16.919	19.023	21.666	23.589
10	12.549	15.987	18.307	20.483	23.209	25.188
11	13.701	17.275	19.675	21.920	24.725	26.757
12	14.845	18.549	21.026	23.337	26.217	28.300
13	15.984	19.812	22.362	24.736	27.688	29.819
14	17.117	21.064	23.685	26.119	29.141	31.319
15	18.245	22.307	24.996	27.488	30.578	32.801
16	19.369	23.542	26.296	28.845	32.000	34.267
17	20.489	24.769	27.587	30.191	33.409	35.718
18	21.605	25.989	28.869	31.526	34.805	37.156
19	22.718	27.204	30.144	32.852	36.191	38.582
20	23.828	28.412	31.410	34.170	37.566	39.997
21	24.935	29.615	32.671	35.479	38.932	41.401
22	26.039	30.813	33.924	36.781	40.289	42.796
23	27.141	32.007	35.172	38.076	41.638	44.181
24	28.241	33.196	36.415	39.364	42.980	45.559
25	29.339	34.382	37.652	40.646	44.314	46.928
26	30.435	35.563	38.885	41.923	45.642	48.290
27	31.528	36.741	40.113	43.195	46.963	49.645
28	32.620	37.916	41.337	44.461	48.278	50.993
29	33.711	39.087	42.557	45.722	49.588	52.336
30	34.800	40.256	43.773	46.979	50.892	53.672
35	40.223	46.059	49.802	53.203	57.342	60.275
40	45.616	51.805	55.758	59.342	63.691	66.766
45	50.985	57.505	61.656	65.410	69.957	73.166
50	56.334	63.167	67.505	71.420	76.154	79.490
55	61.665	68.796	73.311	77.380	82.292	85.749
60	66.981	74.397	79.082	83.298	88.379	91.952
65	72.285	79.973	84.821	89.177	94.422	98.105
70	77.577	85.527	90.531	95.023	100.425	104.215
75	82.858	91.061	96.217	100.839	106.393	110.286
80	88.130	96.578	101.879	106.629	112.329	116.321
85	93.394	102.079	107.522	112.393	118.236	122.325
90	98.650	107.565	113.145	118.136	124.116	128.299
95	103.899	113.038	118.752	123.858	129.973	134.247
100	109.141	118.498	124.342	129.561	135.807	140.169
120	130.055	140.233	146.567	152.211	158.950	163.648

Source: Odeh et al., 1977.

Table A.2 Cumulative Standard Normal Distribution

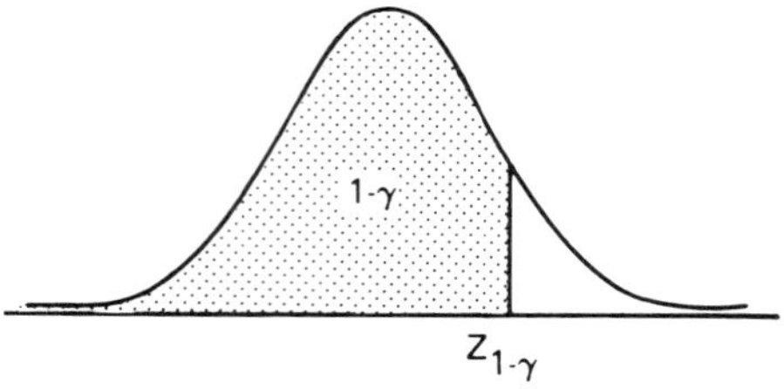

Z	.00	.01	.02	.03	.04	.05	.06	.07	.08	.09
0.0	.5000	.5040	.5080	.5120	.5160	.5199	.5239	.5279	.5319	.5359
0.1	.5398	.5438	.5478	.5517	.5557	.5596	.5636	.5675	.5714	.5753
0.2	.5793	.5832	.5871	.5910	.5948	.5987	.6026	.6064	.6103	.6141
0.3	.6179	.6217	.6255	.6293	.6331	.6368	.6406	.6443	.6480	.6517
0.4	.6554	.6591	.6628	.6664	.6700	.6736	.6772	.6808	.6844	.6879
0.5	.6915	.6950	.6985	.7019	.7054	.7088	.7123	.7157	.7190	.7224
0.6	.7257	.7291	.7324	.7357	.7389	.7422	.7454	.7486	.7517	.7549
0.7	.7580	.7611	.7642	.7673	.7704	.7734	.7764	.7794	.7823	.7852
0.8	.7881	.7910	.7939	.7967	.7995	.8023	.8051	.8078	.8106	.8133
0.9	.8159	.8186	.8212	.8238	.8264	.8289	.8315	.8340	.8365	.8389
1.0	.8413	.8438	.8461	.8485	.8508	.8531	.8554	.8577	.8599	.8621
1.1	.8643	.8665	.8686	.8708	.8729	.8749	.8770	.8790	.8810	.8830
1.2	.8849	.8869	.8888	.8907	.8925	.8944	.8962	.8980	.8997	.9015
1.3	.9032	.9049	.9066	.9082	.9099	.9115	.9131	.9147	.9162	.9177
1.4	.9182	.9207	.9222	.9236	.9251	.9265	.9279	.9292	.9306	.9319
1.5	.9332	.9345	.9357	.9370	.9382	.9394	.9406	.9418	.9429	.9441
1.6	.9452	.9463	.9474	.9484	.9495	.9505	.9515	.9525	.9535	.9545
1.7	.9554	.9564	.9573	.9582	.9591	.9599	.9608	.9616	.9625	.9633
1.8	.9641	.9649	.9656	.9664	.9671	.9678	.9686	.9693	.9699	.9706
1.9	.9713	.9719	.9726	.9732	.9738	.9744	.9750	.9756	.9761	.9767
2.0	.9772	.9778	.9783	.9788	.9793	.9798	.9803	.9808	.9812	.9817
2.1	.9821	.9826	.9830	.9834	.9838	.9842	.9846	.9850	.9854	.9857
2.2	.9861	.9864	.9868	.9871	.9875	.9878	.9881	.9884	.9887	.9890
2.3	.9893	.9896	.9898	.9901	.9904	.9906	.9909	.9911	.9913	.9916
2.4	.9918	.9920	.9922	.9925	.9927	.9929	.9931	.9932	.9934	.9936
2.5	.9938	.9940	.9941	.9943	.9945	.9946	.9948	.9949	.9951	.9952
2.6	.9953	.9955	.9956	.9957	.9959	.9960	.9961	.9962	.9963	.9964
2.7	.9965	.9966	.9967	.9968	.9969	.9970	.9971	.9972	.9973	.9974
2.8	.9974	.9975	.9976	.9977	.9977	.9978	.9979	.9979	.9980	.9981
2.9	.9981	.9982	.9982	.9983	.9984	.9984	.9985	.9985	.9986	.9986
3.0	.9987	.9987	.9987	.9988	.9988	.9989	.9989	.9989	.9990	.9990
3.1	.9990	.9991	.9991	.9991	.9992	.9992	.9992	.9992	.9993	.9993
3.2	.9993	.9993	.9994	.9994	.9994	.9994	.9994	.9995	.9995	.9995
3.3	.9995	.9995	.9995	.9996	.9996	.9996	.9996	.9996	.9996	.9997
3.4	.9997	.9997	.9997	.9997	.9997	.9997	.9997	.9997	.9997	.9998
3.5	.9998	.9998	.9998	.9998	.9998	.9998	.9998	.9998	.9998	.9998
3.6	.9998	.9998	.9999	.9999	.9999	.9999	.9999	.9999	.9999	.9999
3.7	.9999	.9999	.9999	.9999	.9999	.9999	.9999	.9999	.9999	.9999
3.8	.9999	.9999	.9999	.9999	.9999	.9999	.9999	.9999	.9999	.9999
3.9	1.0000	1.0000	1.0000	1.0000	1.0000	1.0000	1.0000	1.0000	1.0000	1.0000

Source: Gunst and Mason, 1980.

Table A.3 Cumulative Student *t* Distribution

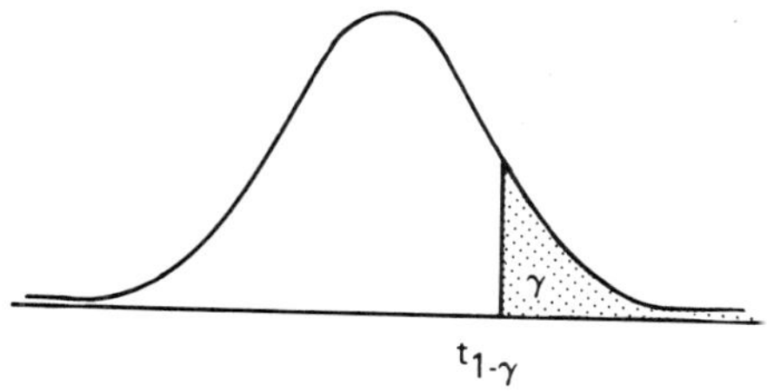

d.f.	$\gamma = .25$	$\gamma = .10$	$\gamma = .05$	$\gamma = .025$	$\gamma = .01$	$\gamma = .005$
1	1.000	3.078	6.314	12.706	31.821	63.657
2	0.816	1.886	2.920	4.303	6.965	9.925
3	.765	1.638	2.353	3.182	4.541	5.841
4	.741	1.533	2.132	2.776	3.747	4.604
5	0.727	1.476	2.015	2.571	3.365	4.032
6	.718	1.440	1.943	2.447	3.143	3.707
7	.711	1.415	1.895	2.365	2.998	3.499
8	.706	1.397	1.860	2.306	2.896	3.355
9	.703	1.383	1.833	2.262	2.821	3.250
10	0.700	1.372	1.812	2.228	2.764	3.169
11	.697	1.363	1.796	2.201	2.718	3.106
12	.695	1.356	1.782	2.179	2.681	3.055
13	.694	1.350	1.771	2.160	2.650	3.012
14	.692	1.345	1.761	2.145	2.624	2.977
15	0.691	1.341	1.753	2.131	2.602	2.947
16	.690	1.337	1.746	2.120	2.583	2.921
17	.689	1.333	1.740	2.110	2.567	2.898
18	.688	1.330	1.734	2.101	2.552	2.878
19	.688	1.328	1.729	2.093	2.539	2.861
20	0.687	1.325	1.725	2.086	2.528	2.845
21	.686	1.323	1.721	2.080	2.518	2.831
22	.686	1.321	1.717	2.074	2.508	2.819
23	.685	1.319	1.714	2.069	2.500	2.807
24	.685	1.318	1.711	2.064	2.492	2.797
25	0.684	1.316	1.708	2.060	2.485	2.787
26	.684	1.315	1.706	2.056	2.479	2.779
27	.684	1.314	1.703	2.052	2.473	2.771
28	.683	1.313	1.701	2.048	2.467	2.763
29	.683	1.311	1.699	2.045	2.462	2.756
30	0.683	1.310	1.697	2.042	2.457	2.750
60	.679	1.296	1.671	2.000	2.390	2.660
90	.678	1.291	1.662	1.987	2.368	2.632
120	.677	1.289	1.658	1.980	2.358	2.617
∞	.674	1.282	1.645	1.960	2.326	2.576

Source: Gunst and Mason, 1980.

Index

A

B

C

D

E